The Mechanics of Human Movement

This book is one of a series on physical education edited by

John E. Kane, M.Ed., Ph.D.

Director of Physical Education,
University of Leeds, and formerly
Head of Department of Physical Education,
University of London Institute of Education,
St Mary's College of Education, Twickenham

Other titles in this series include

Exercise Physiology by Vaughan Thomas
Curriculum Development in Physical Education edited by
John E. Kane

The Mechanics of Human Movement

B. J. Hopper, M.Sc.

Crosby Lockwood Staples London

Granada Publishing Limited
First published in Great Britain 1973 by
Crosby Lockwood Staples
Frogmore St Albans Herts
and 3 Upper James Street London W1R 4BP

ISBN 0 258 96896 6 (hardback)
     0 258 96907 5 (paperback)

Filmset in Photon Times 12 pt. by
Richard Clay (The Chaucer Press), Ltd., Bungay, Suffolk
and printed in Great Britain by
Fletcher & Son, Ltd., Norwich

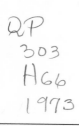

Contents

Foreword

Human movement studies have become increasingly attractive in recent years to undergraduate and graduate students. The attraction is no doubt associated with the identification and clarification of the sub-areas which together currently constitute the broad-based study of human movement. Philosophical approaches have helped to define the pertinent areas of knowledge on which the study draws for structure and relevance. Sociological interpretations have given rise to a clearer understanding of the social and cultural function of play, games, and sport. A wide use of psychological knowledge has, of course, continually been made in the study of human movement in so far as, for example, skills learning and performance motivation have always been central issues, and of recent years the psychology of sport has become a serious academic discipline. The consideration of human movement as an art form has taken on new perspectives which have encouraged a more profound study of the qualitative and aesthetic aspects of movement behaviour. We have also to recognise that amorphous and multi-disciplinary approach to the scientific analysis of movement known as biomechanics, which incorporates mechanical, morphological, and functional interpretations.

One of the main problems associated with this expansion of

knowledge and understanding of the human movement para-
meters has been the disparate nature of the relevant information. The student has found it necessary to cross a number
of disciplinary frontiers to become properly informed, and
even then the choice and applicability of the information has
not always been obvious. What seems necessary as a first step
is an approach, starting from and being centrally concerned
with the human movement phenomenon, which seeks the
fullest understanding of the art and science of movement
through the application of appropriate analytical approaches
from related disciplines. Only when such a practical multi-
disciplinary analysis has been undertaken can we hope to
appreciate the synthesis and significance of human movement.

Bernard Hopper is one of the few men who have attempted
both the analysis and the synthesis. He has carefully
observed and studied various forms (mostly athletic) of
human movement, and has brought to this study the analytical
tools of the physicist and the applied mathematician. In particular he has developed over the past twenty years a
dynamics of human movement which has made a very special
contribution to physical education and sport. Not only has he
been concerned with research and experimentation in this
area, but he has taken time and care to set down his ideas in
such a way as to make them interesting as well as understand-
able to the student of human movement. In this book there are
a few indications of the special teaching equipment and illus-
trations which he has structured, but the systematic arrange-
ment of the book and the clarity with which the critical
issues are elaborated are a clear reflection of his professional
art as a teacher which is appreciated so much by his students
in Britain and North America.

The book represents an important and long awaited contribution to the literature of physical education and sport for
which we are most grateful.

J. E. Kane

Introduction

It is probably true to say that the acquisition of skill in any form of human activity does not normally proceed from the direct application of known laws, but rather from experience gained by observation of phenomena which make their presence felt during the action itself. Subsequent practice should then leave the performer with ever-increasing empirical knowledge of his activity and of sound technique in its execution, until he may justly be regarded as an expert. The fact that this knowledge has no scientific basis should not detract in any way from the means taken to secure it; for such means are recognised elements of scientific method, and should embody more than is implied by mere 'trial and error'. Nevertheless, it would be unfortunate if any human skill were to be allowed to remain indefinitely on an empirical basis. The worthiness of the things we do is judged, not only on their own character, but on our ability to justify their inclusion in the mass of systematised knowledge that we call 'science'; and this is done most effectively when we can point to well-known laws and explain our activity in terms of them.

In these modern times, increasing knowledge in the fields of physiology and psychology, coupled with sophisticated methods of technique-analysis, is making it possible to produce not only a more proficient performer but one capable of taking a more intelligent interest in his form of skill. The importance of this, from the standpoint of mechanics, is that a powerful incentive to the study of the subject has now arisen in a field in which none existed before: a field in which a knowledge of mechanics will become increasingly important.

Considerations of this kind have prompted the compilation of this book. Based on the author's *The Dynamical Basis of Physical Movement* (1959), its general tone is dictated by the need to be intelligible to all interested readers, particularly those with little knowledge of the subject but with the ability to gain more from an appeal to personal experience and common sense than from a strictly technical approach. The student of physical education will find that although the material is treated throughout in the context of his subject, detailed analysis of specific skills is confined to those offering the more obvious illustration of mechanical principles. It is hoped, though, that sufficient insight will be gained from these examples to enable the reader to apply the principles as and where they arise; and, in particular, to investigate with confidence the thought-provoking problems which appear at the conclusion of each chapter.

B. J. Hopper

University of Guelph, Ontario

CHAPTER ONE

Fundamental Ideas

Terminology

Any attempt to describe and explain in simple terms the phenomena which make up the content of an unfamiliar subject, depends for its success on the achievement of a common interpretation of the many observations and experiences with which the exponent of the subject and his readers are concerned: a result which must also involve the universal abandonment of loose, imprecise phraseology in favour of more accurate expression of scientific fact.

In mechanics, the problem is basically one of terminology, for perhaps no other subject lends itself more to the early unfounded acceptance of ill-conceived and erroneous notions; and few suffer more from indiscriminate misuse of their technical terms in the language of common speech. Thus, concepts like those of force, impulse, energy and power all tend to be linked generically in the mind of the layman with the hazy idea of 'effort': something associated in the physical sense with fatigue; while terms designed to describe or discriminate between different types of motion can take on a variety of meanings, and are by no means understood in the same way by all.

An allied difficulty is a literary one—one of communication; for a verbal exposition containing all the elements of precision may be so cumbersome as to be rejected as unintelligible by the reader in search of simplicity. A judicious mixture of precision and approximation is often required, therefore: precision, where principles are involved; approximation, when some deviation from the facts does not invalidate either the discussion or the final conclusion. The 'point of contact' of the foot of a high-jumper with the ground, for example, is not really a 'point' at all, but a small area where the surface is seen to be disturbed; but if the variable pressure of the foot over this area, or the rotational effect achieved as a result of it, is an irrelevant or negligible factor, then it may justly and more conveniently be treated as a mathematical 'point'.

Particles and real bodies

The conflict between clarity and accuracy is well illustrated in the case of a cricket ball (or baseball), which can be regarded as a *point* in any discussion in which its dimensions introduce no pertinent factors; but such discussion would then be limited to matters far removed from the practical conditions of the game itself, in which aerodynamic forces and other effects profoundly modify the path of the ball and force us to regard it as a piece of matter—a body—having finite dimensions and important surface irregularities.

The complication we introduce whenever we abandon the notion of an ideal 'particle' for something real is seen again when we consider what we mean by the 'speed' of a ball. The speed of a particle—an indefinitely small body—is obviously the rate at which it traverses distance along its path; but that of the ball is much less definite since a unique value of speed in a given direction can seldom be ascribed to any part of it except the 'particle' at its centre. If we take its speed to be that of its centre, we are ignoring the fact that the speed and direction of the impact of its *surface* with the ground, or with a bat, is not the same as this when the ball is rotating; nor is the result the same.

Laws of Motion

It is the business of mechanics to formulate principles which enable us to reduce the complex movements of real bodies to their elements, and so to make them amenable to more simple treatment; and the system that does this effectively in the world of everyday things (as distinct from that of sub-atomic and cosmic events) is based on the three statements known as Newton's Laws of Motion. These are statements summing up common experience on matters concerning force and motion, and are justified in that a self-consistent scheme of mechanics has successfully been built on them.

The First Law

> If a piece of matter is at rest it will stay at rest; and if it is moving with a constant speed in a straight line it will continue to do so, as long as no external force acts on it.

From this it follows that:

> If a piece of matter is not experiencing the action of external force it is either at rest or moving with constant speed in a straight line.

In terms such as these the First Law may be expressed so as easily to be appreciated by the layman. That it is of fundamental importance is seen from the fact that it is the basis of a definition of *force*, the agency responsible for change in a body's state of rest or of uniform (steady) speed in a straight line. If change of this sort is not taking place, the body must either be completely free from the action of any single external force or be acted upon by a system of forces which are cancelling each other out.

Effect of extraneous forces

Now although discussion of a philosophical character would certainly reveal difficulties of interpretation of the law and the need for further thought on the precise definition of some of the terms used to express it (it is clear, for example, that it only applies to material 'particles'), yet at much lower levels it might well be considered difficult to find a statement more easily understood than this. Matter and force are constantly part of our experience, as are states of rest and uniform motion in a straight line, and we have no difficulty in thinking of changes in these conditions; yet our failure to understand dynamical phenomena can usually be traced to lack of faith in the validity of Newton's First Law. The reason for this is, surely, that the operation of the Law in the world of common reality has always been masked for us by two serious disturbing factors—the effect of weight and the effect of friction and other incidental resistances to motion.

No one, in fact, normally thinks of a body at rest unless it is being supported all the time by a force exerted on it in an upward direction by whatever it is 'resting' on: a force which we accept as equal in magnitude and opposite in direction to the body's weight (we accept this wholeheartedly when we use a pair of weighing scales). Similarly, uniform motion is never possible, even on a horizontal surface, without the action of a *driving force* big enough to overcome whatever frictional or other resistances to motion there may be. This means that our experience is seldom, if ever, the experience of a single force acting on a body, but of a combination of forces, e.g. a driving force and a resistance, only one of which may be holding our attention. Few of us have ever for very long experienced conditions of apparent 'weightlessness' with no appreciable resistance: conditions ideal for demonstrating Newton's Laws, and achieved temporarily by gymnasts bouncing from a trampoline and by high-divers going into a pool; but these are experiences rare enough and of such short duration as to be popularly regarded more as violation of physical law than in any way confirming it. However, recent television transmissions from spacecraft have convincingly shown the truth of the First Law, and have shown also that we shall not always be handicapped in our approach to matters concerning it, but will easily be able to appreciate the fundamental quality of 'sluggishness' or 'inertia' which material bodies exhibit: inertia which is responsible for maintaining constancy of their speed in magnitude and direction when no external forces are acting.

Weight and mass

Our present difficulty in this matter may well be illustrated by considering our mental attitude towards the 16-lb shot as we attempt to change its state of rest as it lies on a smooth horizontal floor. An upward pull of anything less than 16 lb wt is found to have not the slightest effect on it, but a very small *horizontal* force will start it moving over the floor, however slowly. We are inclined, therefore, to recognise its 'heaviness' but to be not so conscious of its 'massiveness', or 'inertia': we regard it as a 'weight' rather than a 'mass'. Experiments with a punch-bag or a sack full of sand hanging from a high fixed support, however, will show convincingly that the tendency of such a body to resist rapid horizontal displacement from rest, or, when set swinging, to go on moving past its normal hanging position against all our efforts to stop it quickly, is a property of its mass—not of its weight, which is virtually removed from the experiments by the support.

Force and acceleration

It is important to realise that the inertial property of matter only shows itself when a body is being accelerated or retarded. It is responsible for the difficulty of 'getting up speed' and of braking quickly to rest; it makes it impossible to impart to a body an *instantaneous* change of speed, for such would require an infinitely big force; and, in this field, well-known techniques such as the rapid, controlled withdrawal of the hands in the catching of a fast-moving ball are used to increase the stopping-time, and so to reduce the rate of retardation and the stopping-force. The same principle is involved in the use of mats and other cushioning devices in gymnasia, for the purpose of reducing landing shocks.

Acceleration, therefore, is always associated with force, but to maintain a body's *constant* speed in a straight line against no resistance requires no applied force at all. This is, presumably, the condition enjoyed by any material body wandering in space far from the action of gravity. Unless this is clearly understood it will be impossible to appreciate the dynamical basis of any form of physical activity, for it is necessary to realise that force is responsible for change in motion, but not for its uniform continuance.

'Nett' external force

Since we live in a world where forces such as weights and resistances are always acting on the bodies we study, the term 'force' often means for us 'nett' or 'resultant' force. Our 16-lb shot, for example, remains at rest on the floor or on a scale-pan, in accordance with the First Law, because the cancelling of its weight by the upthrust of its support leaves it under the action of no 'nett' or 'resultant' force.

Forces acting on a body are designated 'external' forces when they are delivered to the body by agencies outside it. 'Internal' forces are mutual actions and reactions between component parts of a body: they have no effect on the motion of the body as a whole, although rotary effects may be produced by them (see Chapter Six).

Resistance to a Body's Motion

Terminal speed: constant driving force

When a body moves through a medium such as air or water, the medium opposes its motion by exerting on it a force, commonly known as 'air resistance', 'water resistance', or 'drag'. This force, which depends on the extent to which the medium is being disturbed, therefore increases considerably as the speed of the body through the medium becomes greater: a fact brought out by the uprising curve OR in the Force–Speed graphs of Fig. 1.

Fig. 1(a) illustrates the motion of a body such as that of a sky-diver dropping vertically from rest, possibly from a stationary balloon, under the action of a constant driving force, i.e. his own weight, represented here by the constant ordinate OP. At first, the whole of this force is available for downward

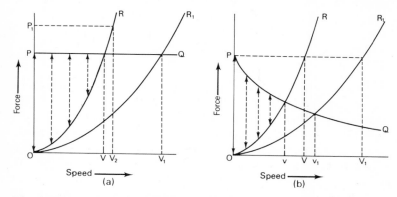

Fig. 1. The attainment of terminal speeds under different conditions.

The arrows show the diminishing value of the *nett* force on a body as the resistance of the surrounding medium increases with speed, and eventually becomes equal to the 'driving' force.

In (a) it is assumed possible for the driving force to be kept constant: in (b) it decreases with speed. (a) has been partially superimposed on (b) for purposes of comparison.

The curves OR and OR$_1$ show the way in which resistance to the body's motion increases with its speed under different 'streamlining' conditions.

propulsion, and the diver accelerates against no resistance; but as the speed increases, so does the air resistance, and the diver continues to reach higher speeds under the action of a nett force (body-weight minus resistance) which the graph shows to be getting progressively smaller. The result is that although the speed goes on increasing, it does so more and more slowly, to reach its 'terminal' value at the limiting speed OV, at which resistance exactly cancels body-weight and the nett force on the body is zero.

The uprising curve OR_1 indicates a reduced disturbance of the medium, the moving body possibly adopting a more 'streamlined' attitude in it. In the case of the sky-diver, this means a delay in the complete cancellation of the driving force until a higher terminal speed OV_1 is reached.

If, by carrying an appreciable load without in any way altering the resistance-characteristics of his fall, the diver could increase his total weight to that shown as OP_1 in Fig. 1(a), then the terminal speed would increase to OV_2: a real but rather insignificant change.

Terminal speed: muscular driving force

In all physical movement, the faster the limbs are made to move, the smaller the 'drive' they can deliver to their surroundings; and the smaller, therefore, is the propulsive reaction the body can obtain from those surroundings. In the case, therefore, of human or animal movement from rest in a resisting medium, the ordinate representing driving-force will not be of constant value as in Fig. 1(a), but will diminish from its maximum, OP, as speed increases; this is shown by the downgraded curve, PQ, in Fig. 1(b). Cancellation of this reduced driving-force therefore takes place earlier than would otherwise be the case—at a speed represented by Ov, rather than OV.

Fig. 1(b) shows the effect to be even more marked when superior streamlining is adopted, the terminal speed, Ov_1, being considerably less than the value OV_1 which a constant driving-force would produce under these conditions. The features of this graph are exemplified particularly in activities like speed-skating and sprinting.

Scalars and Vectors

Speed and velocity

In common speech the terms 'speed' and 'velocity' have virtually the same connotation—rate at which distance is being covered—and are measured in the same units, e.g. ft/s, km/h, etc; and so far there has been no ambiguity in the use of these terms. However, in mechanics a distinction is drawn between them, for whereas 'speed' is regarded as the distance travelled by a body in unit time *along the path it is taking*, no matter how irregular this may be, the term 'velocity' is reserved for the rate at which the body is changing its position *in a given direction*.

The performance of a cross-country runner, for example, depends on accurate judgment of the speed with which the different parts of the course should be covered: a judgment which is determined by the nature of the terrain—not by the direction to be taken. Throughout the run he will be travelling at various speeds in different directions, and will therefore, at every instant, have a particular speed in a given direction; i.e. he will always have an 'instantaneous' velocity, varying with his rate of movement and with his direction of movement.

Direction of motion, as well as rate of motion, is an essential feature of velocity; whereas rate of motion alone is needed to specify speed. Our expression for the First Law of Motion could well be modified so that the term 'velocity' replaced 'speed in a straight line'; for in this form the statement would make it clear that force is responsible for change of velocity: whether by changing a body's speed (leaving its direction of motion unchanged), or by changing its direction of motion (leaving its speed unchanged), or by changing simultaneously both its speed and its direction of motion.

The distinction drawn between speed and velocity is one that illustrates the difference between physical quantities, known as *scalars*, which can be described completely by their magnitude, and others, called *vectors*, which not only have this attribute but that of direction as well. Apart from speed, scalars are represented by such things as mass, energy, and power, all of which have magnitude with no directional features whatever. Vectors, which can be differentiated from each other by direction as well as by magnitude, are exemplified by velocity, acceleration, force, and even by the position of a body with respect to a given fixed point or origin: a position specified simply by the length and direction of the straight line joining the body to the origin.

Vectors: geometrical constructions

The fact that a straight line of finite length is itself a vector makes it possible to represent all vectors diagrammatically by means of such lines, for to do so it is only necessary to draw the representative line in the direction of the given vector and to make its length proportional to the magnitude of the vector. This, and other important features of vector quantities, can be illustrated by a practical example: that of a shot-putt at the instant of release of the missile in a direction which, let us say, bisects the angle between the horizontal and the vertical (Fig. 2). If we assume the missile's velocity to be 45 ft/s, then the straight line **OA**, which represents this, will lie in the required direction, and be 45 units long, each unit signifying 1 ft/s. The line **OA** should also carry an arrow to indicate the positive direction of the vector it represents, as in Fig. 2, and the plane of the diagram should also be stated unless it is already obvious.

One of the fundamental characteristics of a vector is shown

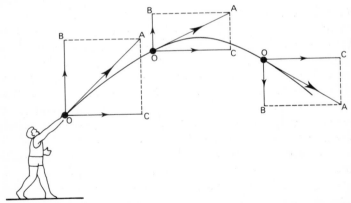

Fig. 2. The changing velocity of a missile is examined in terms of its horizontal and vertically-upward components. The former stays virtually constant: the latter is steadily reduced and eventually reversed by the effect of the missile's weight. At the high-point of the path, **OB** is zero.

by the fact that although **OA** truly represents the velocity at the instant of release—i.e. it describes this velocity completely *by itself*—yet there are other vectorial lines associated with it which we can draw if we need further information. It may be pointed out, for example, that the shot is moving in more than one direction at the same time: it is changing its position in the vertical direction, for it is moving higher; and in the horizontal direction also, since it is moving to the right in Fig. 2. These 'component' velocities must themselves be represented by lines such as **OB** and **OC** in the appropriate directions; and these must be of such lengths as to carry the missile vertically as high as the point A, and, also, as far to the right as this point. Their lengths must, in this diagram, form the adjacent sides of a square, therefore; and each will be found to be about 32 units long, representing vertical and horizontal velocities of 32 ft/s.

'Resolution' of vectors

In examples like that of Fig. 2 it is very profitable to 'resolve' the velocity-vector into its vertical and horizontal components, for these are much easier to study than the velocity-vector itself. It is easy to see, for instance, that with air resistance negligible compared with the weight of the shot, the horizontal component, **OC**, must stay constant throughout the motion; but the vertical one, **OB**, will be steadily reduced by the downward force of the missile's weight, first to zero (at the high-point of its path) and then to increasing values in the downward direction as its flight progresses. Vector diagrams illustrating such changes are shown in Fig. 2. Even without prior knowledge we might expect the constant weight of the shot to produce, as it does, a constant rate of change of the vector **OB**, i.e. a constant downward acceleration.

Addition of vectors

The geometrical figures showing the relation between a vector and its vertical and horizontal components must evidently be either squares or rectangles; but these are special cases, for a vector can be resolved into components lying in any direction with respect to it, the resulting figure being, therefore, a parallelogram.

The converse process of combining, or adding together, two vectors to discover what their 'resultant' is—i.e. what effect they have when they act together—makes use of the same construction. It is clear, for example, from Fig. 2, that the two vectors **OB** and **OC** give rise to their resultant, **OA**; and that this is represented by the diagonal of the figure which meets these lines at O.

Polygon construction   It should also be clear that, in adding the two vectors to obtain their resultant, only half the figure need be drawn, **BA** being used instead of **OC**. This leads to the method for adding more than two vectors, illustrated in Fig. 3. Here, we suppose, an analysis of upper-body movement is being made during a shot-putting action; **OX** represents the velocity of the shoulder with respect to the ground, **XY** the velocity of the wrist with respect to the shoulder, and **YZ** that of the shot with respect to the wrist in the final 'hand

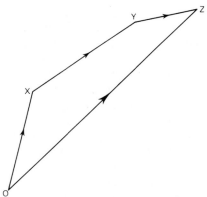

Fig. 3. Vector quantities are added geometrically by arranging their representative lines, such as **OX**, **XY**, **YZ**, in cyclic order from an origin, O, to their extremity, Z. **OZ** then represents, in magnitude and direction, the resultant obtained by the summation.

and finger-flip' stage of the putt. The resultant velocity of the shot with respect to the ground (**OZ**) is obtained by aligning the three components as shown, preserving their true lengths and directions, and ensuring that the arrows associated with them follow the figure from O to Z. Note that although this 'polygon' is here assumed to be drawn in one plane—the plane of the paper—the component vectors will, in practice, not necessarily be confined to it, and the figure will usually be three-dimensional.

'Free Fall'

One important effect of weight has already been noted, namely, the need to apply an upward force on a stationary body, at least equal to its weight, in order to support it or to give it motion in an upward direction. Another effect, curious and seemingly paradoxical, is concerned with the motion of all bodies allowed to fall freely—i.e. in the absence of any resistance—under their own weight. Such bodies, heavy or light, all experience, at the same distance from the earth, the same rate of increase of velocity: an increase which at ground level is about 32 ft/s for every second of their fall. This means that, in 'free fall', all bodies will be travelling with the same velocity after dropping from rest for the same time or through the same distance.

This phenomenon can only be due to the fact that the weight of a body—the force of attraction which the earth has on it—is proportional to its inertia. Although the force impelling a heavier body downwards is greater, so too, and in the same proportion, is its mass; so we find, with slight variations over the earth's surface, the same downward acceleration to be experienced by all freely-falling bodies, and the identical rate of decrease of velocity (retardation) to be experienced by all bodies projected upwards. We do not find the 12-lb shot travelling further than the 16-lb one projected with the same velocity from the same height. With negligible air resistance the two throws would use identical paths and would take equal times. On the other hand, a shot-putt made at a place where gravitational force at the earth's surface is below average (as it is in places like India, Australia, and Japan) could well be an inch or two longer than an identical effort made elsewhere.

Velocity and acceleration We have spoken of a body's increasing speed in the downward direction, and of the equally familiar loss of speed experienced by it when it is moving upwards under the same conditions of freedom; both effects have been ascribed to the same cause, namely, the body's weight. However, since the latter is the only force responsible for these changes of speed, no matter which way the body is moving, it is obviously more satisfactory to identify the direction of the body's acceleration with the direction of the force, and to recognise that all bodies moving freely under the force of gravity are always being accelerated downwards, no matter in what *directions* they may be moving. This only means recognising that a retardation in the upward direction is the same thing as an equal acceleration downwards; or, that retardation is negative acceleration. We assert, therefore, that all bodies moving freely under gravity near the earth, no matter what their direction of motion may be, experience this common vertically-downward acceleration of about 32 feet per second every second (ft/s^2), or 9·8 metres per second every second (m/s^2). This is the factor responsible for the steady change in the magnitude of the vertical velocity-vector of Fig. 2: a change which first brings it to zero and then gives it increasing magnitude in the downward direction.

At this stage it is convenient to recognise the important fact that a body's velocity and its acceleration are entirely independent of each other, even as far as their directions are concerned, unless special conditions exist. The implication is that in studying the mechanical features of a body's motion, velocity is not all that has to be known: this is, in any case, determined by forces that have already acted on it. What is usually more important is its present acceleration, for this is determined entirely by the resultant force acting on it *now*.

Another implication concerns the use of the term 'free fall', which is now a misnomer, for it should denote the condition, not only of 'falling' bodies, but of all bodies moving freely in any way at all under the action of gravity alone. Such bodies have velocities in a multitude of different directions, but their *accelerations* will all be vertically downward, and, near the earth's surface, will all have values very close to the one quoted above: the value always indicated by the symbol g.

Apparent weightlessness

We can now consider certain differences between the dynamical condition of the human body standing at rest on the ground, and therefore bearing its weight on its feet, and the same body in the same attitude rising or falling freely in the air—possibly from a trampoline. Since, in the first case, the body is not accelerating, the force of gravity acting on it (its weight) is being cancelled by the equal and opposite upward force exerted on it by the ground. This force is referred to as 'normal ground-reaction', and in this case is simply supporting the whole body. Now this cancellation of weight is not a condition existing only between the body and the ground; for if the body is divided by imaginary horizontal sections, the weight of everything above a given section must be supported by everything below it, and the increasing extent to which such support is required at individual levels from the head downwards depends on the mass-distribution in the body. It is obvious, in fact, that the standing body is under a degree of compression which reaches a maximum at the soles of the feet, which 'take' all its weight.

All these mutual forces between contiguous parts of the body, due to their weight, are absent in the case of a body in 'free fall' (e.g. a body in the air over a trampoline). We have already seen that once a body has been projected from its support it immediately partakes of the downward acceleration, g, which all such bodies have, and that this causes it to reach a high-point and then to drop back to lower levels with ever-increasing speed. One corollary to this is that not only do separate bodies experience this common acceleration, but different parts of the same body do as well. This, in turn, means that the human body moving freely under gravity not only lacks support for itself as a whole, but for its individual parts also. The supporting forces between horizontal sections no longer exist, the state of compression is no longer there, and the feet no longer 'take' any of the weight; as far as the body is concerned, all its components are weightless, and objects carried by it are also weightless. Such objects can be discarded and allowed to complete their journey unassisted, to be reclaimed in the same relative position with respect to the body before the landing is made. During flight, the body can be held in postures which would be exhausting under static, ground-contact conditions.

Trampoline experiments

The conditions of apparent weightlessness discussed above can be exploited to some extent to gain experience of phenomena due to inertia only. As we have seen, in free fall over a trampoline no effort has to be made by the body to support the weights of its separate components: a phenomenon ascribed to the common downward acceleration possessed by all parts of the system. On the other hand, if the arms are swung rapidly forwards in the early part of the 'airborne' phase of a bounce, and then quickly stopped in a horizontal or upwardly-inclined position, muscular effort will be found necessary both to start the movement and to stop it, especially if dumb-bells are carried. At the same time, the attitude of the rest of the body will be found to have changed from the erect position to one forwardly-inclined.

The forces needed to initiate such relative motion of one part of the system with respect to another, and then to halt it, are due to the inertia of the parts concerned: not to their weight. They are required because relative motion is started only by the *acceleration* of one component with respect to another, and halted only when this acceleration is reversed. Under free fall conditions they are not required to *support* any of the components concerned. The forces are 'internal' forces resulting from muscular contraction, and act equally but in opposite directions on those parts of the body they cause to move. A relative change of position of one part of the body, as in the above example, must therefore change the attitude of the rest in the opposite sense. The attitude and posture of the whole body can most easily be restored, before landing, by returning the arms to their original position *by the same path*.

It is important to notice that, in accordance with the First Law, these internal 'actions' and 'reactions' between components can have no effect on the motion of the body as a whole; the precise significance of this point is one of the fundamental questions dealt with later.

Motion in a curved path

It has already been made clear (p. 11) that force is not only responsible for a body's change of speed in the direction in which it is already moving, but is also needed to cause a moving body to deviate from a straight line—to acquire speed in a direction in which none existed before. The simplest form of such motion is evidently that in which the body is made to move in a circular path at constant speed; for, under these conditions, not only is the speed constant but the rate of change of direction is also constant. This suggests that the force responsible for the motion is directed always towards the centre of the body's circular path, and must be of constant magnitude. Such a force would constitute a steady pull on the body, always at right-angles to the direction in which the body is moving, and so would have no tendency to increase or decrease its speed along the circumference of the circle. This is the 'centripetal' force of which the hammer-thrower is well aware when he has his hammer 'circling' around at high speed: the force he has to exert on the handle.

However, in hammer-throwing, as in other activities involving motion in a curve, the speed of the moving body does not remain constant—it increases with time. This means that, even if the ball traverses a truly circular path, the wire

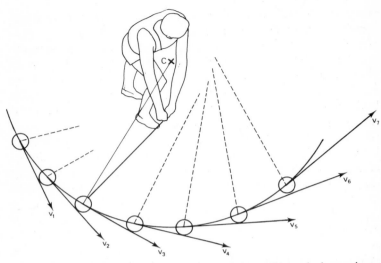

Fig. 4. The motion of a 'hammer' at increasing speed in a circle requires a tangential component of force as well as a centripetal one directed towards C. The tangential component gives rise to a series of increasing velocity-vectors, v_1, v_2, etc., while the radial one keeps the motion going in a circular path. The resultant must always be directed along the flexible wire. (The weight of the hammer has not been taken into account here.)

along which the force acts will not be directed exactly towards the centre, C, of the circle, but will be forwardly-inclined as shown in Fig. 4. The pull in this forwardly-inclined wire can be resolved into two components: the centripetal component acting towards C, and maintaining the circular character of the motion; and the tangential component which accelerates the ball in the direction in which it is moving. This means that the instantaneous velocities shown as vectors in Fig. 4 not only change direction as the motion proceeds, but increase in magnitude also.

'Action' and 'reaction'

The impossibility of initiating the *action* of a force on a body without at the same time evoking an equal and opposite *reaction* from another one is something that constantly claims our attention in the study of human movement: it is a principle that is, in mechanics, of universal application.

Exponents of all forms of skill should be aware of the many ways in which it enters into their activity. Opposing tug-of-war teams, for example, would agree that when a state of static balance has been achieved between them, the pull they are exerting on the rope is equal to the unyielding 'resistance' to motion which the latter is providing in the opposite direction. Lack of motion is, in fact, ascribed to this resistance. Moreover, each member is well aware of the contribution he is making, in the backward direction, by the forward thrust he is exerting against the ground; and he knows and feels that the origin of his contribution is the backwardly-inclined ground-reaction delivered to him in the form of a compressive thrust through his trunk and legs. All the forces are *pairs* of forces acting on components of the whole system; and it is a consequence of the First Law of Motion that if any component is being held at rest, as in the present example, or if it is moving with no acceleration, the nett force acting on it must be zero; i.e. the 'pairs' must be pairs of equal and opposite forces, generally referred to as 'actions' and 'reactions'.

Centripetal and centrifugal forces   The forces involved in steady motion in a circle are common examples of equal and opposite forces. Centripetal force (centre-seeking force) is the force exerted *on* the body moving in the circular path, and directed *towards* the centre of that path. Centrifugal force (centre-fleeing force) is the reaction to this: the force exerted *by* the moving body on whatever is holding it to its circular path. It is directed *away from* the centre of the motion, tending to pull this agency from the centre.

It is important to realise that *centrifugal force does not act on the moving body*.

Problems

1. You have difficulty in stopping a fast-moving cricket ball, travelling horizontally. Consider why you would have the same degree of difficulty if exactly the same incident could happen on the Moon, where the ball would weigh less than an ounce.

2. Explain why a game of cricket, played under 'Moon-gravity' conditions, but with the ball made up to have the normal regulation weight, would be played with no fast bowlers.

3. Why does a weight-lifter, in the early stages of a lift, have to exert on the loaded bar a force greater than its weight? Why is normal ground-reaction at this stage greater than the combined weight of the lifter and the load?

4. A weight-lifter is exerting an upward force on the bar equal to twice its weight. Show that its upward acceleration is equal to g.

5. Refer to Fig. 1 to justify the pronounced crouching posture adopted by competitors in a downhill ski-race and in speed-skating. Explain why the more 'massive' skiers might have an advantage in their event. Consider whether extra mass is helpful or not to the skater (a) in the accelerating phase and (b) in the top-speed phase of a race.

6. Find, with reference to Fig. 3, the velocity of the shot with respect to the shoulder, if its velocity with respect to the ground is 35 ft/s.

7. Show, from Newton's First Law, that a tug-of-war team which pulls its opponents over the line with uniform velocity is exerting no greater force on the rope than they are.

CHAPTER TWO

The Turning Effect of a Force

Elements of a lever

The lever in its many forms is so familiar a device that most people have acquired from it a fairly clear idea of the factors which determine the rotary power of a force. The action of the crowbar and wrench provide ample evidence of the 'shifting power' of a comparatively small force applied at the end of a long handle; while the application of force to the pedal of a bicycle is soon found to be most effective when delivered at right-angles to the crank.

In practice, we recognise lever-action by the existence of a fixed pivot, fulcrum, or axis about which the whole system can be made to rotate by the action of a force, often known as the 'effort'. The 'effort' (or 'power') is applied so that the line along which it is acting does not pass through the axis; i.e. the vector-line representing the effort is eccentric to the axis, and it is opposed by the contrary turning-effect of a resistance, or 'load', also acting eccentrically. Possible dispositions of effort, axis, and load are shown diagrammatically in Fig. 6 (p. 29).

Rotary effects and the parallelogram law

Experience of the way the relevant factors determine the rotary effect of a force is summed up by the statement that its turning 'moment' about a point is given by multiplying its magnitude by the perpendicular distance from the point to the line along which the force is acting. This product is often known as the 'torque': it is the product of two vectors—force and distance—the two vectors being at right-angles to each other and producing (or tending to produce) a rotary effect about the given point. The full value of this torque is itself a vector, for the rotation it produces is about an axis, real or imaginary, drawn through the given point at right-angles to both force and distance-vectors; i.e. perpendicular to the plane of the diagram (Fig. 6). There is some confusion over the units in which torque should be expressed, for if the force is expressed in lb wt and the distance in feet, the product will be in lb wt-ft; but this is commonly shortened to lb-ft.

Consider the static equilibrium of forces shown in Fig. 5. The weight of the swimmer (W lb wt) at the end of the board has a turning-moment about the edge of the bath, E. This is tending to rotate both him and the board in the clockwise sense about E, the magnitude of the torque being the product of W and the length EA; for EA is drawn horizontally from E to meet the vertical line of action of the man's weight *at right angles*. Suppose the board to be held at its other end, S, by a flexible cable, ST, having the direction shown; and suppose the tension in this cable to be F lb wt. Then it is clear that the force F also has a turning moment about E: this time in the anticlockwise sense, and measured by the product of F and the length EB. If, at this stage, we ignore the weight of the board, then we can say that the system will remain static, with no movement around E in one sense or the other, when these two turning-moments or torques cancel each other out, leaving no nett, or resultant, turning-moment about E.

There is, however, another way of expressing this condition; namely, by saying that the resultant of the two forces concerned, W and F, must *itself* have no turning-moment about E, and must therefore have its line of action passing *through* E. This fact makes it possible to incorporate a vector-diagram into Fig. 5: one in which **OP** represents the weight, W, and **OQ** the tension, F, in the cable; and the relative

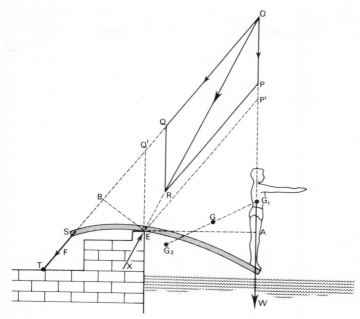

Fig. 5. The turning-effects of static forces about a fixed axis through E are related here to the vector-diagram in which the resultant of these forces is directed through E.

The scale of the vector-diagram can be adjusted, for convenience, to make E and R coincide so that Q is displaced to Q′, and P to P′.

lengths of these lines is determined by the fact that the diagonal of the parallelogram drawn with them as adjacent sides must be directed towards E. Given, then, the weight of the swimmer, the geometry of the figure enables us to find the magnitude and direction of the force F (proportional to the length OQ), and the magnitude and direction of the thrust on the pivot E (proportional to the length OR). The *reaction* at E, i.e. the force X which maintains E in a fixed position, is equal and opposite to the resultant **OR**.

Torque represented by an area   The basis of the concept of 'turning moment', or 'torque', can now be seen in the parallelogram law; for if we adjust the scale of the vector-diagram to make R coincide with E—and this can be done without interfering with any principle—the moment of W about E ($W \times$ EA) is seen to be represented by twice the area of the triangle OP′E; and the equal and opposite moment of F about E ($F \times$ EB) is likewise seen in twice the area of the equal triangle OQ′E. Note that each of these (enlarged) triangles has half the area of the parallelogram itself, and is the product of half one of its sides (**OP′** or **OQ′**) and the appropriate distance (EA or EB) perpendicular to it. Thus the condition that the parallelogram shall have its diagonal, **OR**, directed through E (so giving rise to no loss of static equilibrium) is also the condition that the turning-moments involved shall be measured by the product of force and *perpendicular* distance from E to its line of action: not by any other distance.

Although E appears to be the most obvious point in Fig. 5 about which rotation of the board could occur, equilibrium, if it is achieved under the action of the forces considered, must be achieved about *all* points of the system. Each end of the board, for example, could be chosen as the pivot, or point of balance, and the appropriate parallelogram of forces drawn for it. 'Levers' can be considered to rotate about any point: not just the obvious one.

Centre of Gravity

Resultant of parallel forces

From a study of situations similar to that of Fig. 5, it can be seen that two conditions are needed for the static equilibrium of a body:

(1) The resultant of all the forces acting on it must be zero; i.e. there must be no 'nett' force giving the body linear motion in any direction.

(2) The resultant of the clockwise and anti-clockwise turn-ing-moments of these forces about any point (and there-

fore about any 'axis') must also be zero. This ensures that there shall be no tendency for the body to start rotating: something that could otherwise happen even if condition (1) were fulfilled.

In Fig. 5, for example, the forces F, W, and the supporting force X fulfil both conditions because X is equal and opposite to the resultant of the other two, and acts along the same line as this resultant; but it often happens that all the forces acting on a body are parallel to each other, although having separate and distinct lines of action, as in the simple lever-diagrams of Fig. 6. Here it is impossible to represent them by the sides of a geometrical figure, although equilibrium conditions must still hold. The resultant of the downward forces P and W, for example, must be cancelled by the upward thrust, $P + W$, of the support at O, and must also act vertically downward through the same point O (Fig. 6(a)).

However, in the study of human movement we are constantly meeting systems of forces more complex than this: parallel forces comprising the multitudinous array of indefinitely small 'weights' of all the particles making up the total weight of each material body. These small forces all have the same downward direction, but their lines of action are distributed throughout the volume of each body, and their resultant—the line of action of the body's weight—must be located so that it has the same turning-moment as theirs, about any point. It will, of course, pass through all points at which the body can be statically balanced: points at which an upward force equal to body-weight would have to be applied to support the body in a particular attitude; and one unique and very important point on this resultant is the one at which it can be balanced *in whatever attitude it may be*. This is known as the 'centre of gravity' of the body, always indicated by the symbol G. It is the one point at which the body can be freely pivoted (if it is accessible) in any attitude, with no tendency to rotate into another one: it is the point at which the whole weight of the body may be supposed to act.

It is now possible to be precise about the line of action,

OG_1, of the swimmer's weight in Fig. 5: it is the vertical line through his centre of gravity, G_1. We can also make the above discussion more realistic by introducing the distributed weight of the board into it: a force arising at its centre of gravity, G_2. If the relative weights of the swimmer and the board are known (they are assumed to be equal in this figure), then the combined weight will have its centre of gravity at G, and its vertical line of action can be drawn through this point.

Levers in the body

The transformation of muscular effort into limb-movement is a process embodying the lever principle, for such movement is essentially of a rotary character, and the muscles of the body have their origins and insertions at places which enable them, in contracting, to rotate the limbs about the joints. Their situation is not one that develops great turning-moment because the insertions are comparatively close to the joints; but rapid movement of the limbs is made possible instead because a small contraction of a muscle located close to a joint is necessarily associated with much greater, and therefore faster, displacement of the far end of the limb.

The three recognised classes of lever, together with examples of their occurrence in the body, are illustrated in Fig. 6. The distinction to be drawn between them is one concerning the relative position of the points of application of the 'effort'; i.e. the force operating the lever, and the 'load' or 'resistance' which the effort has to *balance* if it is to hold the lever-arm steady, and to *overcome* if it is to start the lever doing useful work. In levers of the first class the lines of action of these forces pass on opposite sides of the axis, or 'fulcrum', and give rise to situations like those of Figs 6(a) and (d); but in levers of the second and third classes they pass on the same side of it, as in the other examples given. It can easily be seen that in the second-class lever, in which the load operates closer to the fulcrum than the effort does, the latter is always smaller than the load it will balance; but in the third-class lever the reverse is the case. Clearly, most of the body-levers are of the third-

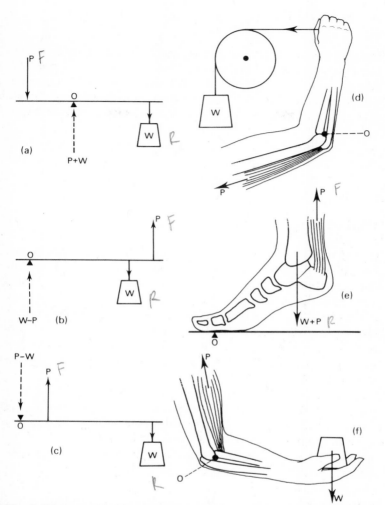

Fig. 6. Diagrams (a), (b), and (c) show the essentials of the three classes of lever. The corresponding diagrams (d), (e), and (f), show examples of these in the human body. P represents the 'power', W the 'load', and the fulcrum is at O. The force exerted by the fulcrum on the lever-arm is shown in (a), (b), and (c).

In (e), W is the vertical thrust of the body on the ankle-joint, together with the weight of the foot itself. P is assumed to act vertically. (W will be one half body-weight if this is taken equally by both feet.)

In matters concerning levers, the terms 'power' and 'effort' are synonymous.

class, the most notable exception being that of the foot, which, in turning about the ankle-joint, can be considered from the standpoint of the body itself to be a lever of the first-class; but from that of the ground making contact with the ball of the foot and the toes, may be regarded as one of the second-class.

Problems

1. Consider how the force-diagram of Fig. 5 would need to be altered to take into account the weight of the board, assumed equal to that of the diver.

2. Reproduce the parallelogram of Fig. 5, and use it to draw others showing that (a) the force F is the resultant of X and W, and (b) W is the resultant of X and F.

3. Assuming static conditions in Figs 6(a), (b), and (c), check that the force exerted, in each case, by the fulcrum on the lever-arm is as stated.

If the effort (power) is not applied vertically in these diagrams, show that it must be greater than P; and that if it is applied at $45°$ to the vertical, then the reaction of the fulcrum must have a component equal to P in the horizontal direction.

4. If, in Fig. 6(e), the whole body is in static balance, explain why its centre of gravity must be vertically above O, and why the downward force on the ankle-joints is greater than body-weight. Why does this weight *appear* to have a line of action displaced from the vertical line through O?

CHAPTER THREE

Forces and their Complete Specification

The Second Law

It is unnecessary to discuss the concept of 'force' beyond pointing out that we appreciate it in the physical world as that on which all material movement depends. The application of force by muscular action is accompanied, we find, by fatigue, and we can make rough comparisons between forces on the basis of this; or, in the case of external forces, by the discomfort or damage experienced by the body when such forces are applied to it. During the application of force to a stationary body, however, it should be obvious that its immediate tendency is to cause motion to start; and when the force is a 'checking' one we see the retarding effect it has on a body already moving. These cases, numerous as they are, should make us realise that the true attribute of force is the capacity it has to give a body acceleration in the direction in which it is acting.

What is required for purposes of measurement, though, is a precise and rational method of making comparison between the magnitudes of forces; and this we get by taking the fundamental effect for which force is responsible, namely, acceleration, and then comparing the accelerations that various forces produce in a given piece of matter; e.g. the standard pound or kilogramme. This intention is implicit in Newton's Second Law, which, for present purposes, may be expressed as follows:

> The rate of increase of velocity, i.e. the acceleration, of a given material body is proportional to the (nett) force acting on it, and takes place in the direction in which the force is acting.

Mass

The above statement has been hinted at already, for it not only identifies the direction of acceleration with that of the applied force—something intuitively taken for granted—but its expression of strict proportionality between the magnitudes of the two quantities implies that the inertia of a material body is independent of its motion, and is therefore intrinsic to the body itself. This independence only breaks down at velocities near that of light. 'Mass', the measure of a body's inertia, is seen here simply as the constant of proportionality between the force applied to a body and the acceleration which that body thereby experiences: the constant appearing in the relation:

$$\text{force} = \text{mass} \times \text{acceleration}$$

or, symbolically,

$$F = m \cdot a \tag{1}$$

Systems and units of measurement

To make calculations based on (1), above, we need a system of units in which the relevant quantities can be measured; the ones commonly in use are based on *length*, *mass*, and *time*. In the English (f.p.s.) system, the foot is the unit of length, the pound is the unit of mass, and the second is the unit of time. The corresponding units in the SI (Système International d'Unités) are the metre, the kilogramme, and the second; these will be more widely used as time goes on.

Reference to equation (1) shows that one unit of force will be the force capable of giving unit acceleration to unit mass: on the f.p.s. system this will be the one giving an acceleration of one foot per second per second to a mass of one pound: it is known as the *poundal*. The SI unit, known as the *newton*, is the force associated with an acceleration of one metre per second per second in a mass of one kilogramme; but, in practice, neither of these 'absolute' units is particularly convenient because we can often get a better idea of how a moving body is behaving if we compare the forces acting on it with its own weight or with the weight of a pound or of a kilogramme.

To make this comparison we remember that when the weight of a body is allowed to act on it alone, i.e. in free fall, it always gives it a downward acceleration g (about 32 ft/s^2 or 9·8 m/s^2); so a force of F lb wt, which gives a body an acceleration a, is related to the weight of the body, W lb wt, by the equation:

$$F/W = a/g \qquad (2)$$

a statement expressing the fact that the ratio of two forces, measured in the same units, is the same as the ratio of the accelerations, also measured in the same units, which those forces produce in the given body. If W is expressed in lb wt, then F will be also: if W is body-weight, then F will be either a fraction or a multiple of body-weight.

As a practical example, suppose we wish to find the horizontal acceleration experienced by the 16-lb shot when a horizontal force of, say, 40 lb wt acts on it. First we remember that a force equal to its own weight (16 lb wt) would give it an acceleration of g; then, since 40 is $2\frac{1}{2}$ times 16, it must give rise to an acceleration of $2\frac{1}{2}$ times g; and if g is taken to be 32 ft/s^2, the result will be 80 ft/s^2.

In equation (2) and examples based on it, the mass of the body does not appear; but cases will be met with later wherein it needs to be preserved. This can be done by referring back to equation (1); and various devices are in use for enabling this equation to be used with forces measured in lb wt or kg wt instead of in the unfamiliar poundals or newtons. The best way to do this is to realise that 1 lb wt is about 32 poundals, and 1 kg wt is about 9·8 newtons. Henceforth, then, *if all forces are expressed in either lb wt or kg wt*, equation (1) will appear as either:

$$32 . F = m . a \qquad \text{in the f.p.s. system}$$

or

$$9 \cdot 8 . F = m . a \qquad \text{in SI}$$

This has the advantage of preserving mass as the single symbol, m, instead of as a ratio. It must always be remembered, though, that the numbers 32 and 9·8, as used in this way, are just pure numbers having no physical significance at all. They are conversion factors used to convert F lb wt to poundals, and F kg wt to newtons, and so to make it possible to use equation (1) in this more convenient way.

Two points should now be noted: one is that the *number* of pounds (or kilogrammes) in the mass of a body is the same as the *number* of lb wt (or kg wt) expressing its weight; the other is that the weight of a body is $m . g$ poundals (or newtons).

Centre of Mass

It is clear from what has gone before that at least two things must be known about a force before its effect on a body can be estimated: its magnitude, which we generally express in either lb wt or kg wt, and its direction. Evidently, its magnitude determines the rate of change of velocity that the force is giving the body—i.e. its acceleration; its direction is that in which this acceleration is taking place.

These two features of a vector quantity have already been discussed in connection with parallelogram and polygon constructions; but there has always been one factor which has either limited the scope of the discussion or introduced an element of unreality into it, and that is concerned with the difference between the ideal notion of a 'particle' and the bulky reality of the material body in which we are generally interested. Whereas it is easy to visualise the precise point of application of a force on a particle (it must be the particle itself), and to recognise the precise character of the motion imparted to it (linear acceleration in the direction of the force), the extended form of a voluminous body makes these things either difficult to specify, or meaningless. Where, for example, is the 'point' of application or the 'line' of action of the force exerted temporarily by a trampoline on the body of a gymnast landing prone in the middle of it; and, if this information were available, how could we use it to decide what the man's subsequent motion must be?

A difficulty of this type has already been dealt with as far as *weight* is concerned, by replacing the distributed weight of an extended body by its resultant: the weight of an equally heavy particle situated at the 'centre of gravity' of the body (a point not located at any fixed part of the body, but changing position as body-posture changes).

Now the Second Law deals with motion—not with a condition of static equilibrium—and the fact that its statement makes no mention of either the line of action of the force or the point at which it is applied can only mean that these factors have no significance as far as the linear acceleration produced by the force is concerned. Our difficulty in applying it to real material bodies is due, therefore, to our present inability to say precisely what is meant by the acceleration, velocity, and even the position of such bodies; particularly when they are of irregular form and non-rigid.

A new concept is required, therefore, to bring extended bodies into conformity with the laws of motion that are obeyed by particles. It must be one which, in our thinking, replaces the distributed *mass* of such a body by that of an equally *massive* particle; and this must be so located with respect to its real associated counterpart that it obeys the laws of motion no matter how the latter comes under the action of force. This is the concept of 'centre of mass', and it is essential that its fundamental attribute be realised—the attribute of moving under all circumstances in accordance with the laws of motion; for it is on this that its rôle in dynamics depends. This is the property that ensures the complete irrelevance, as far as linear motion is concerned, of the complex configuration of the human body and its varying distribution of mass; so that, no matter what complicated movements the subject may have performed since starting from rest, the velocity of his centre of mass depends only on the 'force-pattern' he has so far experienced; and its present acceleration has a magnitude and direction determined by the nett force acting on the body *now*, having a value completely independent of how or where the force is being applied. In particular, when the body is moving in free fall, its centre of mass has the downward acceleration common to all massive particles in this condition, and it maintains this acceleration as it moves in its smooth parabolic path, irrespective of any changes of either attitude or posture which the body may undergo.

Identity of centre of gravity and centre of mass

It is easy to see that for bodies of the size we are considering, centre of gravity and centre of mass coincide; for just as all bodies in free fall experience the same downward acceleration, so do all parts of the same body under the action of their own individual weights. As we have seen earlier (p. 17), this results in the condition of apparent weightlessness between component parts of such a body: a condition in which no part exerts a vertical force on any other part, and the body therefore shows no tendency to rotate into another attitude as it moves. Its weight has no turning-moment about its centre of mass as the latter traverses its path in accordance with the laws of motion; no 'lever-arm' exists between centre of gravity and centre of mass, and the two points therefore coincide at *G*. This is only another way of saying that the distribution of weight is the same as the distribution of mass.

Direct and eccentric forces

The fact that a body's weight has no tendency to start the body rotating shows that its centre of gravity—the origin through which the resultant weight always acts—and its centre of mass share the same point, G. The weight, W, therefore, is a 'direct' force capable of being balanced out when the body is supported at G, without rotation, by another equal and opposite force. The important thing is that *all* direct forces or combinations of forces having their resultant directed through G will, similarly, have no turning effect about G and will cause no rotary effect on the body.

'Eccentric' forces are distinguished from 'direct' ones in that their lines of action do not pass through G: they always have turning-moments about G, and, *while giving G the same acceleration as if they were direct forces*, they start the body rotating about it.

Practical demonstrations   To satisfy ourselves that G has the properties ascribed to it we can have recourse to theory or experiment, or to a combination of both. The demonstration of the greatest significance is probably one which makes use of an irregular piece of card or thin plywood sheet, the centre of gravity of which has first been found by balancing it carefully on the head of a pin or similar support. This point is marked with a red circular spot, and similar black ones are marked in concentric circles around it. When the card is sent spinning into the air, rotating rapidly in its own plane as it goes, it is easy to see that the red spot is the one describing a smooth curve (part of a parabola) as the others move in circles around it.

The dynamical condition of such a body is indicated in Fig. 7, in which the arrows show the varying magnitudes and directions of the 'centrifugal' forces exerted *on* the spinning card by some of its elementary particles as they move around G. The forces are all directed away from G, and all tend to move G away from its place at the centre of the rotation. G, the one point which has no local acceleration, is therefore the one point at which all these forces cancel out. These are all internal forces, becoming greater for the more remote parts of the card, and they have no effect on the motion of the body as a whole. This is true in the sense that they have no effect on

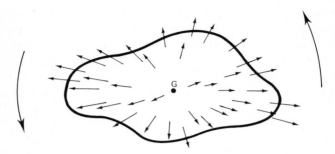

Fig. 7. The multitude of centrifugal forces exerted on a freely-spinning body by its own individual particles have no effect on the motion of G: they cancel at G, which behaves as a massive particle continuing to move, in free fall, in a smooth parabolic path.

The curved arrows show the body's rotation in the plane of the diagram.

the motion of G, do not accelerate it or displace it in any way whatever.

It should now be clear that the concept of a concentration of mass at G, the centre of mass of a body, enables us to reduce the complex mass-distribution of an extended body to the simple status of a particle, and to bring such bodies into conformity with the laws of motion as far as linear motion is concerned.

Yo-yo experiments   To show the behaviour of the centre of mass of a body under a constant eccentric force we can either indulge in a 'thought experiment' or carry out a practical one using a body designed in the form of a yo-yo: a body which can just be prevented from accelerating vertically downwards by a steady upward force delivered to it via its ever-more-rapidly unwinding string (Fig. 8).

It can be argued that the vertical force exerted on the device by the string is doing two things at once: it is preventing its downward acceleration, and is therefore supporting its weight; and because it is causing rotation about G at a faster and faster rate, it is providing a continuous turning-effect without linear movement. Now, to support the weight, a force must be equal and opposite to the weight, and must act through G; and the provision of a turning-effect without linear movement requires a pair of equal and opposite forces having their lines of action laterally displaced from each other, like those commonly used in turning on a tap. It appears, then, that the pull of the string has the same 'lifting-power' as if the yo-yo were doing nothing but hanging passively from it—i.e. it is a force equal and opposite to the weight—and the rotary acceleration of the device is only what we should expect from the eccentric nature of this same force. We conclude, therefore, that a single eccentric force is equivalent to an equal 'direct' one acting through G, together with a pair of equal and opposite forces—commonly known as a 'couple'.

In this example, the forces referred to are each equal to the weight of the body being supported, and will remain so as long as the body is given no vertical acceleration. It is evident

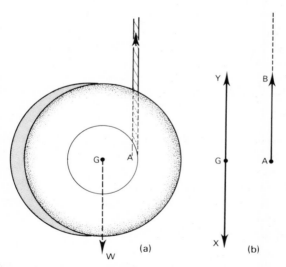

Fig. 8. In (a), a yo-yo is supported by an eccentric vertical pull in its string. It will stay in this position as long as this upward force, shown in (b) as the vector **AB**, is equal to W. (b) also shows **AB** to be equivalent to the three equal and parallel forces **AB**, **GX**, and **GY**. While **GY** supports the body's weight, the 'couple', **AB** and **GX**, gives it rotary acceleration about its central axis.

that if the experiment were to be carried out in the horizontal direction, with the yo-yo resting on a frictionless surface, then the same string-tension would cause the above conditions to be replaced horizontally instead of vertically, with the device accelerating in the direction of the pull in the string at the rate g (about 32 ft/s^2) while its rotary motion speeded up at the same rate as it did before. The use of an experimental body in this form has been described because its rotation has no effect on the relative position of G and the line of action of the eccentric force: in nearly all other cases the eccentricity only remains approximately constant while the force acts for a very short time—when the body is given a short, sharp 'impulse', in fact.

Eccentric forces in human movement

Although the above argument for the identical behaviour of the centre of mass of a body under the action of equal direct and eccentric forces may be convincingly supported by experiment, nevertheless there are some simple (but very imprecise) demonstrations using muscular effort which do not seem to do so. If, for instance, a short rod (e.g. a pencil) is balanced horizontally on the finger and then projected verti-cally, it will rise and fall in the air without rotation: a result consistent with the action of a short-lived impulse directed through G; but if, from this same horizontal position, it is projected upwards by what is judged to be an identical impulse delivered an inch or so away from G, then its centre of mass rises vertically, as expected, while the pencil rotates around it; but it will not rise as far as it did before. Further-more, if a point of projection near the end of the pencil is chosen, it may even be impossible to raise the centre of mass at all, although a very rapid rate of rotation will be apparent.

Tests such as the one described could raise doubt in the mind of the experimenter about the ability of G to represent the extended form of the pencil when an eccentric force is applied to it; for if G is to fulfil its fundamental function it must gain the same velocity (and in this instance rise to the same height) when a force of given magnitude and direction acts for the same time at *any* point of the body associated with it. The difficulty is only resolved when we realise the great difficulty of applying, physically, a given impulse to an external body under eccentric conditions. Fig. 9 shows that, for this to happen, the agency responsible (the finger) has to be accelerated faster than G, and therefore has to move faster than it would move in delivering a direct impulse to the body.

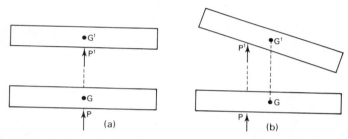

Fig. 9. The contrasting effects of a direct force, P, and an equal eccentric one, acting for the same time, on a rigid body. In (a) the point of application of the force has moved through the same distance (from P to P') as G (from G to G'). In (b) it has moved further than G as the body rotates.

This is a matter having great significance in physical movement: a matter concerning the big difference between forces made available internally by muscular contraction for the purpose of accelerating a limb, and the force which this same limb is capable of exerting simultaneously on an external object. Several factors appear to be involved in this: firstly, the physiological fact that the force made available by muscular contraction becomes less as the speed of contraction becomes greater; secondly, the increase of internal resistance to limb-movement, which also increases with speed; and, thirdly, as in the present example, the necessary rapidity of movement when the external force is an eccentric one: a factor which not only reduces the available force, but reduces also the time for which it is able to act.

It would appear, then, that the development of eccentric force in a physical activity must be accompanied by a limitation in performance; and this would apply, not only to cases such as that cited above, but to the derivation of a force such as ground-reaction, which is usually exerted eccentrically on the body of the performer himself. If this is so, then the disappointing result is not due to the failure of G to obey the laws of motion, but to difficulties of a physiological nature which the performer has been unable to overcome. However, the non-rigid body of a skilled exponent of a movement is well equipped to deal with such difficulties by bringing more than one extending limb into the action at the right time.

The principles involved in this are dealt with later (p. 147) but we may already appreciate the advantage of some degree of controlled flexibility to a body subject to an eccentric force, by studying the behaviour of the device shown in Fig. 10: a jointed rod which can be regarded as a development of the rigid one of Fig. 9. Here a portion of the rod, shown shaded, is cut out and hinged to the main part at H. It is assumed that a release mechanism (not shown) operates as soon as an eccentric force is applied to the body at P, so that the hinged part is allowed to rotate about H in the clockwise sense by the action of a spring, shown at S, which forces it round. In (a) the hinged part is comparatively short, and the spring not particularly strong; in (b) and (c) this part is longer, and the spring assumed to be increasingly more powerful. The initial position of the rod is horizontal in all diagrams, and identical impulses are assumed to be delivered at P, causing identical upward movement of G while they act. In practice, P could well be the upward force of ground reaction exerted on the body via an extending leg.

Now although the application of an eccentric impulse must give a body rotation about G as well as moving G itself (see Fig. 9), the flexibility possessed by the jointed rod of Fig. 10(a) makes it possible for this rotation to be confined to the shaded portion only, while the rest retains its horizontal attitude. If this happens, then the point of application of the impulse moves through a distance PP' which is slightly *less*, rather than greater, than the distance GG' through which G is moved. This result is exploited even more in the remaining

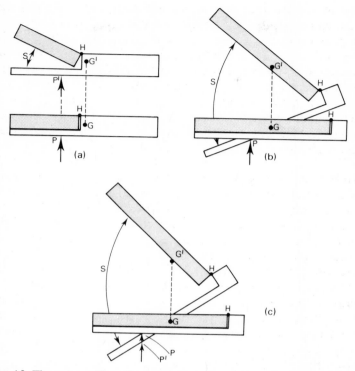

Fig. 10. The rotary effect of an eccentric force can be confined to the part of the jointed body shown shaded in (a), so that the point of application of the force, *P*, moves somewhat slower than *G*.

In (b), the operative spring, *S*, is strong enough to prevent *P* from moving at all; and in (c) this point is even forced back from *P* to *P′*.

diagrams, where, as in (b), it is conceivable that the greater mass and length of the hinged part, operated by a stronger spring, will keep *P* *static* while *G* rises (a condition corresponding to that in which the muscles of the driving leg merely have to hold their position isometrically, and, in doing so, remain capable of exerting a very much greater force until they are able to move *G* even further and faster by their own subsequent contraction). An even more extreme case is shown in (c), where the point of application of the impulse is actually forced *back* from P to P' (simulating the 'stretching' of the muscles of the driving leg before they are able to make their own positive contribution to the motion of G).

It appears, then, that the factors which completely specify a force, and which therefore describe its effect in any form of physical activity, are its *magnitude* and *direction*, its *line of action* and *point of application*, together with the *time* for which it is able to act. Skill in such activity depends on the way these attributes of force are used to produce the desired motion of the body: motion which is most conveniently described in terms of the motion of the body's centre of mass and the rotation of the whole body around it.

Problems

1. The ends of a bar-bell are given different loads. Show that, by exerting lifting-forces on it only, a blindfolded weight-lifter could not detect the uneven distribution of its weight.

To exert equal lifting-force by each arm, where would his hand-grips need to be in relation to the system's centre of gravity?

Explain how it is possible for these forces to be equal when the loads on each side of him are different.

2. Try the experiment of problem 1, above, with a really *flexible* bar: explain the result achieved.

3. Use Fig. 9 to explain why, when an eccentric force has to be applied to a missile (as in discus-throwing), it should be the *final* action in the throw, and made by the arm-extremity only.

4. Find the centre of mass of a rigid rod; then, holding it loosely near its upper end, quickly displace it horizontally. Explain (a) why G moves in the same direction as the sud-

denly displaced support, and (b) how the rotary effect of the eccentric impulse delivered at the support causes some point below *G* to remain, momentarily, at rest.

Find this point more accurately by *shaking* the support rather than merely displacing it; and then find where to hold the rod so that the shaking of it has no effect at its lower end.

Do the same with an inverted cricket bat, and so find where the ball should make contact with it so as to cause no 'jarring' effect in the handle.

(The point of delivery of the blow is called the 'centre of percussion'; the point not immediately affected by it is the 'centre of suspension'.)

CHAPTER FOUR

Friction and Stability

Resistance to motion

In discussing friction in the context of human movement, we have to distinguish between what is called 'sliding' friction—the force which, acting equally and in opposite directions along each of two surfaces in contact, eventually reduces their relative motion to zero—and resistance offered to the passage of a solid body moving through a fluid medium: something already discussed in connection with terminal velocity (p. 8). Whereas the first of these effects is generally essential for stability and for progress over the ground, because it involves 'resistance' to sliding, the second is always wasteful of energy, and therefore unhelpful.

Air resistance   The fact that sprinting performance, in particular, is much less impressive when undertaken against a head-wind is evidence that air resistance has an appreciable effect on the human athlete in rapid motion; the figure of about 4 lb wt against a man travelling in still air at 20 mph or so—the one usually quoted—is the equivalent of a run up a slope of about 1 : 40. Against a head-wind, the resistance will clearly be greater, as it will even when the wind is blowing *across* the track. It is also well known that the wind has an adverse effect on runners making complete circuits of a closed track; for the 'tail-wind' experienced in one section of it has a much smaller effect in reducing air resistance than the inevitable head-wind has, elsewhere, in increasing it. Study of the resistance curves of Fig. 1 (p. 9) will show why this is so; and consideration of the fact that these athletes must spend a higher proportion of their time running at reduced speed against fast-moving air will show the disadvantage of performing under these conditions.

Sliding friction   Consideration of Newton's First Law shows the impossibility of imparting motion to a body at rest except by the action of a force in the direction in which motion is desired. Horizontal speed can only be gained, for example, by a man standing on level ground if a horizontal force acts on him for a finite time. In the absence of any other agency capable of applying it, the necessary force must come from the reaction of the ground; and this reaction must, therefore, have a horizontal component as well as the vertical one supporting the man's weight. In order to move from rest, he will thus have to evoke from the ground a horizontal reaction by exerting an equal force on it in the direction opposite to the one in which he wishes to start moving; and this will usually be done by the forward *acceleration* of one leg—out of contact with the ground—so that the other foot is urged backwards. If there were no sliding friction between this foot and the ground, there would be no resistance to the sliding of it in the backward direction, no forward reaction from the slippery surface, and therefore no horizontal movement of the man's centre of mass. Progress from rest (i.e. acceleration) is thus impossible without the necessary reaction from the ground: if the foot is not 'anchored' to it by spikes, then sliding friction must be responsible for this reaction. Should the foot be urged back with a force greater than the maximum frictional force available, then there will be slipping of the foot backwards, the forward force developed by the ground being equal only to that made available by friction. This maximum frictional force between surfaces in contact is commonly known as 'limiting' friction: it is somewhat greater than the friction between them when sliding has started. Clearly, for purposes of rapid acceleration over the ground it is desirable for limiting friction between the foot and the ground to be as great as possible: this is achieved by the use of suitable surface materials, and by making normal ground-reaction as great as possible.

Balance—its maintenance and recovery

A rigid body standing on a base of finite size is 'stable' provided that the vertical line drawn through its centre of gravity passes within that base; for the weight, acting as a resultant vertical force along this line, has a turning-moment

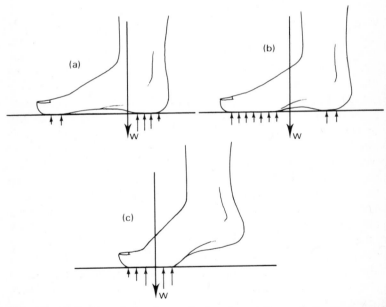

Fig. 11. Variation of thrust on the bearing-surfaces of the foot. The distribution of the small components of normal ground-reaction changes progressively from (a) to (c) as the line of action of body-weight, W, moves forward.

about the edge of the base which will always tend to restore the body to its original position if it is rocked slightly about that edge. It is clear that this restoring tendency becomes less as the displacement about the edge becomes greater, until a condition of instability arises when the centre of gravity, G, is exactly over the edge. Further displacement, however slight, must result in the body toppling to a new position.

In the case of the non-rigid human body standing on a rough floor, the situation is in many respects different from the above; for whereas a rigid body has to be displaced from its stable position by some external agency before it can start to topple, the human body standing erect is already unstable in the sense that it needs the constant involuntary variation of pressure over the surfaces of the feet to maintain its stance. Normally, balance depends on G never being allowed to move to a position which, with a rigid body, would cause instability, although it is never actually at rest. Fig. 11 shows, by the number and by the length of the lines representing 'ground-reaction vectors', how this reaction can be distributed over the bearing-surface of the foot in order to control the line of action of resultant body-weight, even if the latter moves so far forward as to force the heels to be raised. It suggests, too, how the forward movement of this line can be arrested by an early redistribution of the upthrust towards the ball of the foot before the line reaches it.

The fact that, for a standing person, the position of G needs to be controlled, emphasises the essential need for enough friction at ground level to make this possible. In the absence of a horizontal component of ground-reaction, any minor lack of coincidence that may arise between the line of action of the resultant vertical upthrust and that of the weight must cause the body to start rotating about G—the feet sliding away as G drops vertically. Anyone who has stood on a very slippery surface will know the suddenness with which this can happen. However, on a rough surface it will be possible to control the position of G with respect to the feet by deriving from the ground the necessary horizontal reaction. Any suitable

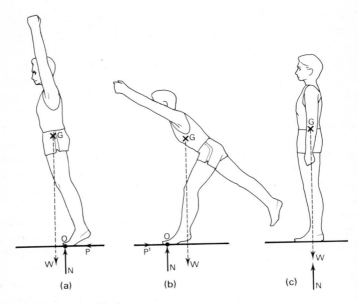

Fig. 12. Recovery of balance.

In (a), balance has been lost. In the absence of friction between the feet and the ground, G would drop vertically as the feet slide back, but the frictional force, P, is preventing this. At the same time, P is urging G forwards.

In (b), rapid rotary acceleration of most of the body in the 'forward' sense forces the supporting foot forward against the backward frictional force, P′, so bringing G back behind the support, O.

Recovery of the erect posture finally forces G over the feet again, as in (c).

posture-change will do this; and, if made vigorously enough, can even be used to gain recovery of balance which, for a rigid body, would be irrecoverable.

Suppose that, as in Fig. 12(a), balance has been lost in the 'forward' direction, G having moved ahead of the toes. Remembering that G responds exactly in accordance with Newton's Laws wherever force is applied to the body, we see that the frictional force required to move it back over the feet must be backward, instead of the forward one, P, that has caused balance to be lost: P, in fact, has to be reversed and made as great as possible in the backward direction. Now it is evident that a horizontal force at ground level is extremely eccentric, and cannot come into operation unless it is associated with increasingly rapid rotary motion of the body (or part of it) about G. So, to develop a force P' in the direction required, and of the necessary magnitude, the posture-change must be similar to the one shown in Fig. 12(b)—an accelerated rotation, in the anticlockwise sense, of all parts of the body that can partake of it: the trunk, the arms, and one leg. When this occurs, the body will be behaving as expected: the eccentric force P' will be giving G acceleration in the backward direction and causing most of the body to accelerate in the rotary sense around it.

Now these parts of the body cannot go on accelerating indefinitely: they must soon be retarded to rest, and their motion then reversed, to regain the erect stance shown in Fig. 12(c). During this period, the direction of P' will change once again: it will become a forwardly-acting force which brings G to rest—ideally, over the feet. A certain amount of skill is needed for this, for the other force shown in Fig. 12—the normal ground-reaction, N—also contributes to rotation about G: first in one sense and then in the other.

Problems

1. Make rough copies of Fig. 12(a) in different 'toppling' attitudes. Show that G's acceleration has two components at right-angles to each other, as with Fig. 4. Note that, as long as the force P has the direction indicated, G's resultant acceleration must be inclined forwards; but will this be the case when the body has nearly reached the horizontal? Show that the feet will tend to slide forward at this stage, especially as N will then be much smaller than W.

Consider the rotary acceleration of the body, and so prove that the resultant of N and P must always be directed up behind G.

2. You are suddenly forced to 'stand on your toes' without first getting G over them. Explain the rapid arm and trunk action you make to preserve balance.

3. Show that, in skating, all horizontal reactions from the ice are virtually at right-angles to the blade of the skate. Point out, therefore, the difficulty of accelerating forwards from rest.

Assuming that air resistance is the only other horizontal force acting on the skater, use the parallelogram construction to show the resultant force acting on him while a driving effort is being made. Show that, at maximum skating speed, this resultant is at right-angles to his forward direction of motion; and discuss, in terms of skating technique generally, the advantage of minimising air resistance.

4. The force, R, exerted by the water on a water-skier, acts on the skis at the 'centre of pressure'. Assuming the pull, P, of the tow-rope to be horizontal, and ignoring air-resistance, show that, at a steady speed, the vertical component of R is equal to W, and the horizontal one is equal to P. Use the parallelogram law to show that the inclination of R depends only on the magnitude of P; and that if this inclination is constant, then the backward lean of the body depends on the line of action of P. Does air resistance make any basic difference to these conclusions?

CHAPTER FIVE

Impulse and Momentum: the Third Law

In what has gone before, emphasis has been laid on the need for a clear distinction between 'accelerated' motion and 'uniform' motion—i.e. between acceleration and velocity—and this has been so because we have been concerned with the action of force—acceleration is the kinetic effect always associated with the action of force. Nevertheless, in practical cases we are not only interested in the immediate result of the application of force to a body but in the final result obtained when the force has been acting for a certain time: we need to know the final velocity of the body—the product of its acceleration and the time for which this acceleration has been going on. One complication which always arises in the study of physical movement is the variable character of the acceleration, both in magnitude and direction, as time goes on: the result of corresponding changes in the force over the same period; but in what immediately follows we shall assume constant values of the quantities involved.

Assuming force, F, to be measured in either lb wt or kg wt, as advocated earlier (p. 34), and using the same notation, we have:

$$32 \cdot F = m \cdot a \quad \text{and} \quad 9 \cdot 8 \cdot F = m \cdot a$$

as statements of the Second Law in f.p.s. and SI units respectively.

To introduce the product of acceleration and time (i.e. velocity) into these equations we multiply throughout by time, t seconds, and obtain:

$$32 . F . t = m . a . t \quad \text{and} \quad 9 \cdot 8 . F . t = m . a . t$$
$$= m . v \qquad\qquad\qquad = m . v$$

The product of force and the time for which it acts ($F . t$ lb wt-s, or kg wt-s) is known as the *impulse*, and the resulting product of mass and gain in velocity ($m . v$ lb-ft/s, or kg-m/s) is the gain of *momentum* of the body to which the impulse has been delivered. It is hardly necessary to point out that impulse and momentum are both vectors associated with each other in accordance with the tenets of the Second Law; and that a force, acting in a given direction for a finite time, will confer on a body momentum in the same direction, irrespective of the momentum already possessed by that body.

Conservation of momentum

The static condition of a body standing on horizontal ground may evidently be ascribed to the cancelling out of the equal and opposite forces of weight and normal ground-reaction. This is in agreement with the First Law of Motion, which denies the existence of any resultant force when there is no acceleration. If we take it that this equality between force and its reaction can be extended to cases where such forces arise by moving bodies coming into contact, or influencing each other by other means, then we are endorsing Newton's Third Law of Motion:

To every action there is an equal and opposite reaction.

During a collision of the kind shown in Fig. 13, for example, between bodies A and B, assumed initially to be travelling in the same direction, with A moving faster than B, the force exerted by A on B in the forward direction is at all times equal to that exerted by B on A in the reverse direction. Not only this, but the forces act simultaneously over an identical time-interval; so it follows that the forward impulse (*gain* of for-

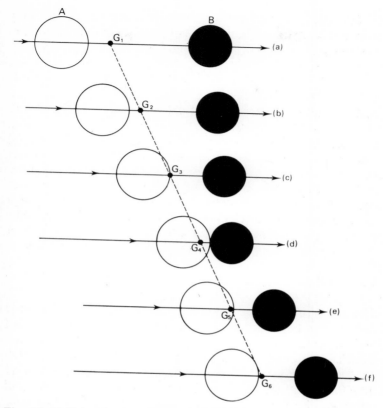

Fig. 13. Collision between bodies leaves the velocity of their common centre of mass unaffected. Their velocities of recession from each other and from G are less than their velocities of approach: an indication of some loss of kinetic energy.

ward momentum) given to B is equal to the backward impulse (*loss* of forward momentum) experienced by A. We thus arrive at the important conclusion that the mutual forces between these bodies have not made the slightest difference to the momentum of the system as a whole; for that gained by one member in a given direction has been lost by the other in the same direction. This law—the Law of Conservation of Momentum—is of universal validity in mechanics, and may be applied to mutual actions between component bodies of any system.

An extension of the First Law   Another way of regarding the interaction between the bodies A and B, above, is to consider the motion of their common centre of mass, G. If, as in Fig. 13, we assume the mass of A to be twice that of B, we see that G must be always twice as far from B as it is from A, and so must move as shown by the points G_1, G_2, G_3, etc., in the separate diagrams until the collision causes A and B to move away from each other, G still maintaining the same relative distance from each of them. The point to note is that although A and B do not have the same speeds as they each had before the collision, their common centre of mass is entirely unaffected by it. This result is clearly an extension of the First Law to systems of bodies and to component parts of the same body, in that no change in the motion of G is seen to be possible if no impulse from outside the system is applied to any member of it.

The rôle of the Earth

It is a little unfortunate that common phraseology has linked the term 'impulse' with an effort that is essentially short-lived; for although many such examples can be cited in the field of physical activity, yet they are by no means exclusive. In fact, as defined, impulse is a quantity which can go on increasing indefinitely with time: always accompanied by another equal and opposite impulse similarly increasing with time.

In the example of static equilibrium of a body standing on the ground, we find that downward momentum is prevented from developing by the action of a contrary impulse increasing with time at the same rate as that due to body-weight: each force, weight or ground-reaction, is called into action by the presence of the other. The example of a freely-falling body, on the other hand, appears to involve only one force—the weight of the falling body itself—but when the origin of this force is considered it is seen that the Earth must enter the argument as well. No impulse, and therefore no increase in momentum, would arise in a region far from the gravitational attraction of a body such as the Earth; and so, if we have to think of this as a two-body system, we have to believe that the force attracting the small body to the Earth is equal and opposite to that pulling the Earth to the small body. It follows that the gain in momentum downwards by the falling body is equal to the gain in upward momentum by the Earth as the two come together; and that when the collision occurs, the equal and opposite momenta are mutually destroyed by the equal and opposite impulses generated on contact.

Although we do not normally think of the Earth as being affected by the attraction of small bodies near it, yet such must be the case if Newton's Laws are valid; for, as with Fig. 13, the two bodies must be treated as a simple isolated system having a common centre of mass: a point incapable of being disturbed by anything happening within the system. Mutual action between two component parts can only result in the motion of *both* parts with respect to *G*: one component cannot stay at rest while the other starts moving. Strictly speaking, therefore, we should make all measurements, not with respect to the surface of the Earth but with respect to the combined centre of mass of the Earth and everything on it and near it. In practice, however, the enormous mass of the Earth ensures that nothing we can do yet has any measurable effect on either its position or its rotation; and we can regard it simply as an indefinitely big reservoir, or source, of momentum: one which remains unaffected by the exchanges of momentum which we make with it.

Heat production within a system of bodies

From the above discussion we should now appreciate that a given impulse is always associated with the appropriate change of momentum, and that internal impulses can never cause change in the total momentum possessed by a system of bodies such as the two spheres of Fig. 13. This momentum, moreover, is represented by that carried by the mass of the whole system, supposedly concentrated at G, and moving as G is moving—unaltered by any interaction between component parts.

Now the collision shown in Fig. 13 between the spheres A and B has necessarily changed their velocities of approach to G into velocities of recession; but there has been another important effect: their velocities of recession from G and from each other are *less* than their velocities of approach. Furthermore, experience shows that if the collisions could be constantly repeated (say, by the provision of a light elastic thread linking them), then A and B would separate with diminishing velocities as more and more contacts were made, until they came together permanently at G. This phenomenon, a decrease in the relative velocities of component parts of an isolated system (no matter by what process it occurs) is accompanied by the production of heat—the bodies become warmer as their relative velocities become less. This is an example of the degradation of energy; energy of motion being changed into the random molecular motion which we know as heat energy.

Problems

1. During a bouncing exercise, a gymnast's centre of mass takes $1 \cdot 0$ s to drop from its high-point on to the bed of a trampoline, and to be brought, momentarily, to rest. Show that the downward impulse on the body due to body-weight is equal to the upward one due to the trampoline; and, if $0 \cdot 3$ s has so far been spent in contact with the latter, show that the mean upward force exerted on the gymnast during this time by the trampoline is $3\frac{1}{3}$ times body-weight.

If, during a long series of bounces, the gymnast spends one-third of the time in contact with the trampoline, show that the upward force experienced by him has a mean value of three times body-weight.

2. Ten ice-hockey pucks are lying close together, at rest, on an ice-rink. Another one is driven hard into the group, causing a succession of collisions as they scatter. Draw a diagram showing the path of the common centre of mass of *all* the pucks, and explain why, in the absence of friction, the magnitude and direction of its velocity remains constant until one puck strikes the boards.

3. Stand on the toes, holding an unloaded bar above the head. Give the bar a sudden forward impulse, moving it a few inches, and then bring it to rest. Explain why the mutual impulses acting between the bar and the body cause the whole system to lose balance backwards.

Try the same experiment with the bar held low, and explain the different result obtained. (Refer to Problem No. 4, p. 44.)

CHAPTER SIX

Rotary Motion

Fundamental ideas

A brief encounter with non-rectilinear motion has already been made: particularly motion in a circle. This is easy enough to describe for a simple body such as a 'hammer' (p. 19) whirling around a fixed axis at the end of its wire, for we can talk in terms of 'revolution-rate' and radius of the circle of revolution.

It is not easy, however, to be precise when the rotary motion is that of the non-rigid human body, capable of rapid redistribution of mass, and of initiating twisting movements in the air which are, apparently, not related to any specific 'axes' of rotation. In order, therefore, to fix our ideas and to define more precisely the terms we intend to use, we start by studying the characteristics of the simplest practical example of rotary motion: that of a circular disc spinning about a material axis through its centre and at right-angles to its surface. A circular piece of stiff card with a sharp pencil stuck through its centre will serve for demonstration purposes (Fig. 14).

The first thing to note is that, whereas linear motion involves change of *position*, the essence of rotary motion is change of *direction*. The point A on the circumference of the disc, for example, will change its position from A to B as the disc rotates about its axis; and in doing so it will travel a linear distance measured by the length of the curved path it takes along the circumference; but the extent of its revolution about the axis will be determined by the change in the *direction* of the radial line OA from its initial direction to that of OB. This is a change indicated by the size of the angle AOB

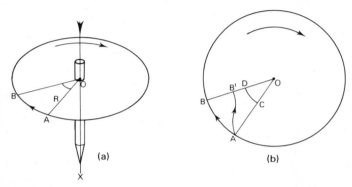

Fig. 14. The fundamentals of rotary motion are illustrated in (a) by a flat disc rotating about its central axis OX. In (b), the angle swept out by all points on the radial line AO are shown to be equal, even if their distance from O changes as the disc rotates.

The direction of the angular velocity vector is, conventionally, that of the arrow drawn along the axis in (a). It is directed away from the observer in (b).

(often called the 'angular' distance) which will continually increase as the disc goes on spinning: and just as the linear velocity of a point is the rate of change of its position in a given direction, so, in the rotary sense, the angular velocity of a point about an axis is the rate of increase of an angle such as AOB in either the clockwise sense, as in Fig. 14, or the anticlockwise.

The introduction of the directional aspect suggests that angular, or rotary, quantities are vectors just as are their linear counterparts; although it is difficult to see how any direction can be ascribed to a line such as OA which is necessarily changing direction all the time. The difficulty is resolved, however, when we realise that the angular velocity of a point about an axis is only completely determined if we know the *direction of the axis* and the sense of the rotary motion going on about it. The direction of the angular velocity vector is, therefore, that of the axis; and its positive sense—the way in which its representative arrow will be drawn—is indicated conventionally by the direction in which an observer must look, along the axis, to see the rotary motion taking place in the clockwise sense.

Units of measurement

Clearly, then, we are required to deal with angular measurements; and we are probably familiar with such units as 'degrees' or even complete revolutions, in such work. However, the fundamental way of defining the measure of an angle is by the ratio of the distances already mentioned—the arc, AB, of the circumference, and the radius OA (Fig. 14). When this ratio is unity, i.e. when the arc has a length equal to the radius, the angle subtended by the arc at the centre, O, is one *radian*. It is well to note that the measure of an angle does not depend on the individual lengths involved in it: the angle AOB in the figure is equally-well defined by the ratio of the arc CD to the radius OC, as by the ratio just quoted. Clearly, then, all points on the rigid disc have the same angular velocity about O, although their *linear* speeds increase from the axis outwards, and are, indeed, proportional to their distances from the axis: a fact which is always true of all parts of a rigid body spinning about an axis fixed in relation to it.

To lead the argument more closely to the case of a non-rigid body, let us suppose that instead of the point A being fixed to the circumference of the disc, it is free to move towards O as the disc rotates (like a fly walking on the surface —not necessarily at a constant speed). A's path might now be the irregular one AB′ (Fig. 14(b)) instead of the circular arc AB; but, even taking this path, it has still swept out the same angle AOB about the axis, so its mean angular velocity during this time is the same as if it had *not* been free to move towards O. Angular velocity depends only on the rate at which the line joining a point to the axis is changing its direction: it is independent of the actual path taken by the point.

Human rotary motion often introduces complication because there seldom exists a fixed point, like O, around which the body is moving in a simple way. Even in hammer-throwing, which is often quoted as a 'simple' illustration of principles, the distance between the massive sphere and the handle can never be regarded as the radius of the curve around which it is travelling, for the hand-grip itself is constantly in motion as the thrower moves about in the circle. The most that can be said is that as the direction of the wire changes, the hammer-head has angular velocity *relative to* its handle: a quantity measured by the rate of change of this direction. The true 'instantaneous' centre of the motion may be more accurately given by the point C in Fig. 4.

A seemingly anomalous case arises when a particle, moving in a straight line (i.e. with motion that appears to lack any rotary features whatever), is seen to have angular velocity (not necessarily constant) about every fixed point not situated directly in its path. In Fig. 15, for example, the point A is shown moving at constant speed along the straight line AZ, covering the equal distances shown in equal intervals of time. Clearly, the line joining A to the fixed point O is changing

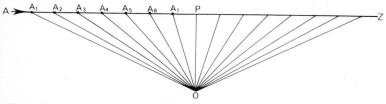

Fig. 15. An observer at O follows the path of a ball travelling along the straight line AZ with uniform velocity. The ball moves through equal linear distances in equal times, but the corresponding angles through which the observer's head has to turn (the angular velocity of the ball about O) increase as the ball approaches P and diminish as it recedes. Even rectilinear motion has rotary features as far as an eccentric point such as O is concerned.

direction in the clockwise sense—it is sweeping out an angle about O at a rate which increases to the point P, and then decreases again as A recedes from it—and is an example of angular acceleration and angular retardation; a practical demonstration of this is provided by the rotary motion of the heads of spectators near the sidelines of a tennis court, as they follow the nearly straight-line path of the ball from one side of the net to the other.

Relation between linear and rotational quantities   Reference to Fig. 14 will show that, just as the angle swept out by a point such as A is measured, in radians, by dividing the length of the arc AB by its radius, R; so the angular velocity is measured, in radians per second, by the linear velocity divided by R. Furthermore, if the point A has angular acceleration, it will be measured, in rad/s^2, by dividing its linear acceleration as it moves round the arc by the same radius, R. Thus, for a rigid body rotating about an axis fixed with respect to it, rotary quantities (i.e. angular distance, angular velocity and angular acceleration of all points on it) are obtained from the corresponding linear quantities by dividing the latter by the appropriate radii.

Dynamics of Rotation

Moment of inertia

We have seen already (p. 54) that a force acting for a given time in a constant direction will give to a material body an increase of velocity (and therefore of linear momentum) in the same direction in accordance with the relation:

$$32 \cdot F \cdot t = m \cdot v \quad \text{or} \quad 9 \cdot 8 \cdot F \cdot t = m \cdot v$$

using the notation of p. 34.

Furthermore, the rotary effect of an eccentric force has been shown to depend on the turning-moment it has about the body's centre of mass (p. 37); and it is reasonable to argue that just as a force, acting for a given time, gives a body linear momentum, so, when a turning-moment acts for a given time, it constitutes what we might call a 'rotary' or angular impulse, and would give the body 'rotary' or angular momentum. The first few vertical bounces of a trampolinist, for example, made in order to gain height for a final somersault, are the result of a succession of impulses directed as accurately as possible through G. The final one, however, must be made *eccentric* with respect to G; and the product of its turning-moment about the transverse axis through G, and the time for which this has operated, is a measure of the angular momentum given to the body about this particular axis. The result is that the centre of mass moves vertically, or nearly so, while the whole body somersaults around it.

Experience clearly shows, then, that matter exhibits the same sluggishness, or reluctance to change its rate of rotation as it does to a change of linear velocity; but the 'rotary' inertia of a body depends not only on its mass but on the way its mass is distributed about the axis of rotation. How difficult it is, for example, to induce a long horizontal pole to turn quickly about one's shoulder, and how difficult to stop the motion quickly, once it has been started! Yet the same pole can be given rotation about its longitudinal axis very easily, and it can be retarded to rest just as quickly. It is not surprising that this should be so; for, in the first case, but certainly not in the second, much of the material is far from the axis about which the pole is being made to turn, and so must be given a high linear speed even when the pole is turning slowly. So it is that different degrees of difficulty are experienced when we try to give angular acceleration to bodies of different shape, or to the same body about different axes. The degree of difficulty, rotary sluggishness, reluctance to change rate of rotation, or however it may popularly be described, is a manifestation of what is known in dynamical terms as 'moment of inertia'; always represented in mathematical terms by the symbol I.

Mathematical relationships

Evidently, moment of inertia is the inertial factor that enters into rotary movement in the same way as mass does in linear movement. Just as the mass of a body determines the linear acceleration that a given force can produce in it, so its moment of inertia about a specific fixed axis determines the angular acceleration produced in it by a given torque (turning-moment) about that axis. The statement: 'Force is equal to the product of mass and acceleration', has its rotary counterpart in the declaration: 'Torque is equal to the product of moment of inertia and angular acceleration'.

It happens that in the study of human movement we seldom have cause to deal with angular acceleration, because the angular *velocity* acquired by the body is of greater relevance; so we realise that, just as linear momentum is the product of mass and linear velocity, the quantity we want is contained in *angular* momentum: the product of moment of inertia and angular velocity. In a similar way we note that angular impulse—the factor responsible for giving a body angular momentum—is the product of *torque* and the time for which it operates: a quantity corresponding to that involving *force* and time in the linear case.

It now remains to express angular impulse and angular momentum in the mathematical symbols commonly used for them, and then to equate them, as was done for their linear counterparts on p. 54, in the units of the f.p.s. system and SI. To do this, we remember that torque, T—the turning-moment of a force, F, about the axis of rotation—is the product of the force and the length (p) of the lever-arm (real or otherwise) at right-angles to which it is acting; and so, multiplying this product by the time (t) for which the torque acts, we get the angular impulse, P, in the form:

$$P = F \cdot p \cdot t$$

To do the same for angular momentum we require moment of inertia, I, multiplied by angular velocity, the usual symbol for which is ω. Angular momentum, sometimes symbolised by J, is thus given by:

$$J = I \cdot \omega$$

We can only equate angular impulse with the angular momentum for which it is responsible if we express force (F) in absolute units. Using the familiar gravitational units, lb wt and kg wt, we apply the appropriate conversion-numbers, 32 or 9·8, as required, to give:

$$32 \cdot F \cdot p \cdot t = I \cdot \omega \qquad \text{in the f.p.s. system}$$

or

$$9 \cdot 8 \cdot F \cdot p \cdot t = I \cdot \omega \qquad \text{in SI}$$

p being measured in feet or metres, respectively, t in seconds, and ω in radians per second. Examination of the dimensions of the quantities involved will show that the appropriate units for I are either lb-ft^2 or kg-m^2: this will be made clear later.

'Remote' and 'Local' Angular Momentum

We are now in a position to study the ways in which the quantities discussed above enter into the behaviour of bodies of various kinds; particularly of the human body in the course of its rotary activities.

1. The case of a particle

First we consider a particle of mass m, at rest at A (Fig. 16) on a perfectly smooth horizontal surface. Let O be a fixed point in the same surface: a point known henceforth as the 'origin'.

Suppose, now, that the particle is set in motion by the action of a short-lived *linear* impulse, $F \cdot t$, in the direction from A to Z; then we know that it will move along AZ with constant velocity v given by:

$$F \cdot t = m \cdot v \qquad\qquad (3)$$

F being expressed in the absolute units of either system.

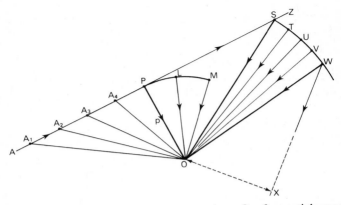

Fig. 16. The constant angular momentum, about O, of a particle moving freely through positions A_1, A_2, A_3, etc., in equal time-intervals, and then along either of the circular arcs shown, is directly related to the sweeping-out of equal areas about O in these times. No change in angular momentum can occur as long as the only force exerted on the particle is directed towards O.

Now, while the force F was acting, it had a turning-moment about O; and to obtain this torque we multiply F by the perpendicular distance from O to the line AZ along which it has acted. The perpendicular line is shown as OP, and has a length p. This tells us that the *angular* impulse about O, which has been delivered to the particle, is simply the linear impulse, $F \cdot t$, multiplied by p (that is, $F \cdot t \cdot p$), and has given rise to angular momentum, $m \cdot v \cdot p$, about O. In fact, we have the very useful result:

$$F \cdot t \cdot p = m \cdot v \cdot p \qquad \text{(}F\text{ in absolute units)}$$

This result shows that the angular momentum of a particle—but not necessarily of any other body—about a given origin is expressible in terms of its mass (m) and two *linear* quantities: its linear velocity (v) and the perpendicular distance (p) from the origin to the line along which the particle is travelling—the line along which the impulse started it moving.

The possibility of linear motion having rotary qualities has been indicated in connection with Fig. 15, where motion with respect to a remote origin is concerned: the possibility of applying this to human movement is seen when we remember that G, the centre of mass of an extended body, can itself be regarded as a massive particle obeying the same laws as any other body; and capable, therefore, of having this 'remote' form of angular momentum about any point not directly in its path.

Geometrical representation of angular momentum   If, in Fig. 16, the successive positions of the particle (A_1, A_2, etc.) are those it has at the end of each second, then each segment (A_1A_2, A_2A_3, etc.) must be of length v, the distance travelled by the particle in these units of time. We then notice that all the triangles having the common vertex, O, have these segments as equal bases, and, with the same common height, p, must therefore have equal areas, each of $\frac{1}{2} . v . p$—areas swept out about the origin O in unit times. The angular momentum, $m . v . p$, of the particle about O can therefore be expressed as the product of twice its mass and the rate at which it is sweeping out an area about O. This link between angular momentum and area is one which also has useful applications.

Constancy of angular momentum of the remote form   There are interesting variations on the motion of the particle A (Fig. 16) which can now be investigated. Suppose, first, that when this particle reaches P it is constrained to move, no longer in a straight line, but in a circular path around O. This could happen if, at P, it suddenly became attached to O by an ideally 'massless' thread of length p. Its motion in the circular path P–L–M, etc., is being caused entirely by the tension in the thread: a force directed always towards the origin, O, and having, therefore, no turning-moment about O. On this account, there is no change at all in the angular momentum of the particle about the origin: it remains at $m . v . p$, showing that the particle continues to move with unchanged speed, v, and continues to sweep out an area about O at the same rate as before.

Angular momentum is evidently independent of the path in

which a particle is constrained to move, as long as the constraint delivers no angular impulse to it about the origin. This is seen with even greater emphasis by considering the situation if the particle had been allowed to proceed along the line AZ to some more distant point such as S before the connecting-thread became taut. If the thread were entirely non-elastic it would suddenly reduce the particle's component of linear velocity in the direction OS to zero, leaving it to move in the circular path S–T–U, etc., at reduced linear speed. Now in spite of this sudden change in the characteristics of its motion —particularly its loss of some linear momentum—there will still be no change in the particle's *angular* momentum about O; for, once again, the sudden impulse changing its path has been directed towards O, and has no torque about O. Again, therefore, the areas OST, OTU, etc., swept out in unit time, are equal to those of the triangles swept out earlier.

If we suppose an elastic thread to be used, or if a sudden 'jerk' is given to it towards O, as is suggested when the particle reaches W, then the same result is achieved. In the latter case, the product of its velocity and the perpendicular distance OX is exactly the same as the initial product of v and p in the early stages of the motion.

We conclude, therefore, that the remote angular momentum of a particle about an origin, O, remains constant unless the particle experiences an impulse having a torque about the origin.

Moment of inertia of a particle about a fixed axis In all that has gone before, the axis about which the motion of the particle has been discussed has been understood to be the one through O, and at right-angles to the plane of the diagram. To find an expression for the moment of inertia of the particle about this axis we can use any part of Fig. 16, for we know its angular momentum to be $m.v.p$ throughout the motion shown there; but the circular path from P, on to L, M, etc., is convenient because here there is constant angular velocity $\omega = v/p$ in a path of constant radius, p. We therefore write:

$$m.v.p = I.\omega$$

and since

$$\omega = v/p$$

$$m \cdot v \cdot p = I \cdot \frac{v}{p}$$

and hence

$$I = m \cdot p^2$$

To generalise: the moment of inertia of a particle of mass m about an axis at a distance r from it is given by:

$$I = m \cdot r^2$$

an expression of the form already quoted (p. 67).

2. The case of an extended body about a 'remote' axis

Fig. 17 is a development of Fig. 16 in the sense that, instead of showing the rectilinear path of a massive particle, it shows

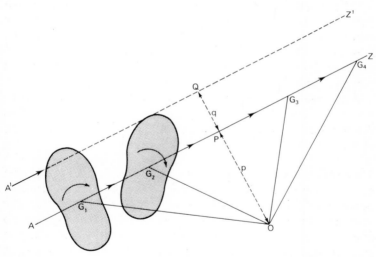

Fig. 17. An eccentric impulse delivered to the extended body along A′Z′ gives G the same linear velocity and the same 'remote' angular momentum as if it had been directed straight through it. In addition, the body has been given 'local' angular momentum, seen in its own rotation about G.

The motion of G along AZ is exactly the same as that of the particle in Fig. 16.

the identical path of the centre of mass, G, of an extended body similarly situated with respect to the origin, O. G could well be the centre of mass of a human body (i.e. a non-rigid one) and, as in the earlier discussion, the motion is taken to be on a perfectly smooth horizontal surface. All relevant axes of rotation and revolution are at right-angles to the plane of the diagram, and that through O is not connected in any way to the extended body: it is 'remote' in the fullest sense.

The dynamics of the situation will be, as we know, identical with that of the particle of Fig. 16 if the same linear impulse $F \cdot t$ is delivered directly through G along the line AZ; for the extended body will then move without rotation, and the velocity of G will be given by equation (3), (p. 67). Its angular momentum about the remote axis through O must also be $m \cdot v \cdot p$, as before, since the same angular impulse has been responsible for it.

Now suppose that, instead of the linear impulse being a direct one, it is delivered in the same direction (i.e. parallel to AZ) but *eccentrically*, so that its line of action does not pass through G. From our recognition of G's fundamental property —its obedience to the Laws of Motion no matter where force is applied to the body associated with it—we see that its velocity will be precisely what it was under the direct impulse: it will move along AZ with velocity v. Moreover, G, treated as a massive particle, carries with it angular momentum about O equal to $m \cdot v \cdot p$, as before. This is the 'remote' angular momentum which the body has about this point: it is in the clockwise sense in Fig. 17.

Now it is evident that this is not the only angular momentum possessed by the body about O, for, in this figure, it has been subjected to an angular impulse greater than that which has set G moving in the way described: greater by an amount $F \cdot t \cdot q$. This excess has obviously been responsible for giving the body angular momentum about G—it is an eccentric impulse giving rise to 'local' angular momentum about the centre of mass of the extended body itself.

From this it is seen that an extended body differs from a particle in that it is capable of rotation about an axis through its own centre of mass, and can possess angular momentum about that axis. Its angular momentum about an axis through a point such as O is not, therefore, just one quantity, $m . v . p$, but the sum (or difference) of two quantities:

(1) Angular momentum due to the motion of G.
(2) Angular momentum due to rotation of the body about an axis through G.

The use of epithets such as 'remote' and 'local' in this connection is purely for convenience. It is constantly necessary to distinguish between these two forms, for, as will be seen later, in many human activities one form can be 'traded' for the other.

3. The case of a rigid body rotating about an axis rigidly connected to it

It is now possible to examine a particular case of Fig. 17: the case where the axis through O is part of the rigid body, and is therefore the one about which the body is constrained to rotate. This situation differs from that already discussed in that G itself is forced to move in a circular path around O, just as everything else is. Moreover, G has angular velocity, ω, about O, which is the same as that of all parts of the rotating body; so the latter's motion can still be described as a rotation about an axis through G, while G moves *at the same angular rate* round a parallel axis through O (Fig. 18).

Choice of expression for angular momentum   The fundamental concept of moment of inertia, I, enables us to express the angular momentum of a body about an axis fixed to it, as:

$$J = I . \omega \qquad \text{(p. 67)}$$

If the axis passes through the point O, as in the present example, then it is convenient to write I as I_O: if it passes through the body's own centre of mass, then this specific

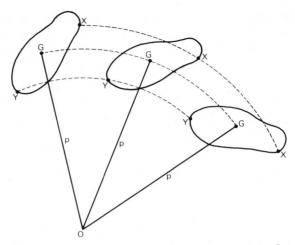

Fig. 18. Here, the body is rigidly attached to the fixed origin, O, so that all parts of it, including G, rotate around O with the same angular velocity. This common angular velocity is also that with which individual points, such as X and Y, are moving around G. Under these conditions the moment of inertia of the body about O is the same as that of G about O, together with the moment of inertia of the body itself about a parallel axis through G.

The parallel axes referred to are, of course, both perpendicular to the plane of the diagram.

value of I is indicated by I_G, and the local angular momentum of the body by the expression $I_G \cdot \omega$.

In Fig. 17 it was convenient to express the remote angular momentum of the body—that due to G's own motion about O—as $m \cdot v \cdot p$; but under the special conditions now being considered, in which G is moving in a circle of radius p instead of in a straight line, the form $m \cdot p^2 \cdot \omega$ is more meaningful. (Note that $m \cdot p^2$ is the moment of inertia of G, treated as a particle, and ω is the angular velocity of the whole body about O.)

We are thus able to write:

$$J = I_G \cdot \omega + m \cdot p^2 \cdot \omega$$
$$= (I_G + m \cdot p^2) \cdot \omega$$

an expression showing that the moment of inertia of such a body about an axis fixed to it is greater than its value about a parallel axis through G by an amount $m \cdot p^2$. This is a statement known as the Theorem of Parallel Axes: it is not usually derived in this way.

In the field of human movement this result shows itself in the comparatively slow pendulum-like swinging of the fully-extended body from the high bar—a rotation in which G is as far from the bar as it can be, and for which the distance p is of the order of $3\frac{1}{2}$ ft. When the necessary effort is made to bring G from near the low-point of the swing to close coincidence with the bar, a much more rapid rotation of the body is generally apparent. Maintenance of an erect attitude on rough ground is made easy by the fact that body-rotation from this attitude takes place about the feet: again, about $3\frac{1}{2}$ ft from G. If, on slippery ground, the feet no longer provide a fixed axis, then rotation takes place about G, and the prone position is reached much more rapidly.

It must be remembered that, although in the majority of activities the body rotates about the *transverse* axis through G, or an axis parallel to it, and has specific moments of inertia about these axes, yet a change of posture which brings the mass of the body closer to them must reduce these values and cause a speed-up of rotation. Furthermore, the moments of inertia of the body about other axes have values which, again, vary widely with posture.

Radius of gyration   We have already met some simplifying artifices such as the concept of centre of mass, which, as far as linear motion is concerned, reduces the distributed mass of an extended body to that of a particle; and we naturally wonder if a similar device can be introduced to help us with rotary motion by replacing the complex mass-distribution of the bodies we have to deal with by something with a simple geometrical form having the same rotary properties. The simple geometrical form would have to be that of a body having the same mass and the same centre of mass as the one it replaces, and having the same moment of inertia about whatever axis through G may happen to concern us.

Many such devices can be suggested, and, from our point of view, the replacement of the distributed mass of the body by an indefinitely-thin circular 'ring' of matter, of mass equal to that of the body, and centred on G, is most satisfactory. This imaginary ring would need to have a radius giving it the same moment of inertia as the body itself has about the particular axis associated with it—the axis through G at right-angles to the plane of the ring. The radius of the equivalent ring is known as the 'radius of gyration' of the body about the axis concerned: it is always given the symbol k.

Principal axes of rotation   Before proceeding further it is necessary to call attention to a phenomenon which drastically reduces the number of axes about which the radius of gyration has to be specified. If any rigid body is made to rotate about its centre of mass, then only three axes can generally be found about which it will do so without tending to force the axis into a different orientation. These three axes are mutually at right-angles and intersect at G. If, for example, a rectangular wooden block has a thin nail driven accurately into the centre of each of its six faces, leaving an inch or two projecting on each side, then the three pairs of 'axes' so formed are axes through G, about each of which the block can be supported and spun without wavering. About any other axis through G, however (e.g. the one from one corner to the opposite one), a spinning motion is accompanied by a pronounced tendency for the axis to tear itself away from its initial direction, and, if free to do so, to gyrate in a more complicated manner in space.

The three axes about which a body will rotate without disturbing the axis in any way are known as 'principal axes' through G, and they are shown for the erect human body in Fig. 19. As will be seen, they correspond with the well-recognised anatomical axes: AB, longitudinal; CD, sagittal; and EF, transverse. The 'rings of gyration' associated with these axes are also shown with their approximate radii, which

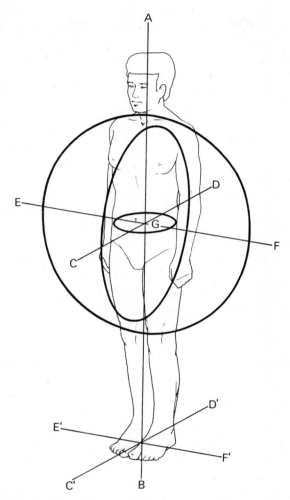

Fig. 19. Principal axes through *G* for the erect human body are shown as
AB, CD, and EF; and 'rings of gyration' associated with these axes are
given their approximate relative radii.

C'D' and E'F', principal axes through the feet, are axes about which the
body's moment of inertia is much greater than it is about any of the axes
through *G* (see Fig. 18).

are nearly the same for the axes CD and EF, about which the erect body's moments of inertia are considerable, but much smaller for AB, about which I_G has only a small value. Principal axes also exist for all other points of the body: those through the feet, parallel to the ones through G, are shown in Fig. 19. The body will rotate about all of these without twisting, but, of course, it will exert a lateral pull on any such axis which does not pass directly through G; e.g. the high bar about which giant circles are being performed.

Properties of principal axes: axis of momentum   It will be realised that if a rigid body is rotating freely in the air about a principal axis, then this axis will be one passing through G, and all parts of the body will be moving in circles with planes at right-angles to it: all the body's local angular momentum is contained in this rotation. This means that the angular momentum is a vector having the same direction as the principal axis, and can conveniently be represented by a line drawn along this axis. The direction of the angular momentum vector and that of any line representing it is known as the 'axis of momentum'; and in the case of a body rotating freely without twist, the axis of momentum must coincide with one of the body's principal axes.

Now it has already been pointed out that, in the absence of any disturbing factor, there can be no change in either the magnitude or direction of a body's angular momentum; so the axis of momentum in free fall retains whatever direction it had on take-off, no matter what postural changes a non-rigid body may undergo. However, a postural change can make a good deal of difference to the direction of the body's principal axes, whether in free fall or not; so it can destroy the coincidence that may exist between one of these axes and the axis of momentum. When this happens, the principal axis about which the rotation was taking place will not only be re-orientated but will not retain a constant direction in space: it 'gyrates', tracing out a conical surface about the axis of momentum (Fig. 20).

Fig. 20 illustrates the change in the characteristics of a

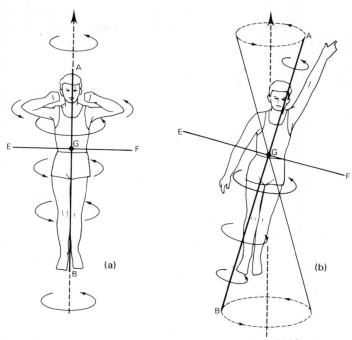

Fig. 20. In (a), the body is spinning in the air about its longitudinal axis, which coincides with the axis of momentum (shown as an upwardly-directed arrow).

In (b), an asymmetrical posture-change has altered the orientation of the longitudinal axis, AB, which now 'gyrates' in a conical manner around the axis of momentum as the body spins. For obvious reasons, this is *not* a practical modification of the ballet-dancer's *tour en l'air*.

simple rotary motion—in this case an aerial pirouette—when a posture-change displaces the principal axis about which the rotation is occurring. In (a), the performer is rotating freely about his longitudinal axis, AB, in the sense shown by the arrows. This principal axis through G is also the axis of momentum, the direction of the angular momentum vector being indicated by the upward arrow. EF is the transverse principal axis. In (b), the asymmetrical positioning of the arms has changed the direction of AB in space so that it no longer coincides with the axis of momentum, which, of course, is unaffected. The body now continues to rotate about the inclined axis AB, but AB itself starts 'gyrating' in a conical fashion about the axis of momentum, tracing out the surface of a cone in the sense shown. This motion of the principal axis is known as 'nutation'—a phenomenon due entirely to lack of coincidence between the axes, and not to any external impulses which may change the direction of the axis of momentum. Such impulses give rise to 'precession' of this axis. (*Note.* Many authorities do not distinguish between 'nutation' and 'precession', and often use these terms to denote any 'wobbling' movement that they happen to be describing.)

Any further posture-change which restores the axis AB to coincidence with the axis of momentum—usually a reversal of the one displacing it—will remove the twisting characteristic of the motion and turn it back into a simple pirouette. Whether the motion is a twisting one or not, the body's local angular momentum is entirely confined to the axis of momentum—it never has any component in the plane at right-angles to this. For convenience, this plane—horizontal in Fig. 20—can be called the plane of 'zero momentum'.

'Stable' and 'unstable' axes   In practice it is impossible to spin a body, rigid or otherwise, *exactly* about one of its principal axes. If an attempt is made to do this with the rectangular block described earlier, or with a model of the erect human body of Fig. 19, by projecting it into the air with rapid rotation about axes AB, CD, and EF, in turn, then the axes AB and CD—those associated with minimum and maximum moment of inertia, respectively—will be found to retain a nearly constant direction in space: nutating, in the way described above, about the axis of momentum, but staying close to it. The transverse axis, EF, however, will show no such tendency, but will swing rapidly back and forth, rhythmically reversing its direction as the rotation proceeds; and this motion will be accompanied by similar violent changes in the direction of the other axes. Only if exact coincidence between EF and the axis of momentum were to be achieved

would this axis retain a constant direction; and this can only be done, in practice, by providing fixed bearings within which EF can freely rotate as an axle. The erect or extended human body, though, with moments of inertia about CD and EF nearly equal, behaves very much as a long cylinder, for which *any* transverse axis is a principal axis. However, in a 'tuck' posture, or the one adopted in clearing a hurdle, moments of inertia of the body about all axes through G are more nearly equal, and it is difficult to specify any particular principal axis. Minor deviation of any principal axis during the body's passage through the air can always be corrected by the appropriate change of posture.

Turntable Demonstrations

Many of the phenomena discussed above can be illustrated by experiments with a 'turntable'. This is a circular platform (sometimes provided with a light stool) which supports the experimenter and whatever apparatus he may be using, so that everything is free to rotate about a vertical axis: the firmly-mounted axle to which the platform is bolted, and round which it moves in a nearly frictionless manner. The experimenter, his apparatus, and the turntable itself form a connected system of bodies sensitive to external rotary impulses about the vertical direction only, and capable of detecting internal mutual actions taking place only in this direction. The platform should have a comparatively small moment of inertia about its axle, and it should be possible to adjust the latter to be truly vertical. It is sometimes convenient to have a small, adjustable amount of friction against the motion of the system, but this is almost an isolated system as far as rotation about the vertical is concerned.

Experiment 1. The effect of a re-distribution of mass   Two dumb-bells, each just capable of being held comfortably at arm's length, are held by the experimenter with arms outstretched as he stands (or sits) centrally on the turntable. The latter is set rotating slowly, its angular momentum remaining sensibly constant. When the two massive bodies are quickly brought in to the neighbourhood of the chest (i.e. close to the axis of rotation) there will be an equally rapid increase in angular velocity: when the dumb-bells are returned to their original, more remote, positions, the angular velocity will be reduced to its former low value.

The well-known explanation for this is that the moment of inertia of the system has been reduced by the greater concentration of its mass near the axis; and since this *internal* change leaves the angular momentum unaffected, an increase in angular velocity has necessarily occurred. The system returns to its initial rate of rotation when the mass-distribution returns to what it was before.

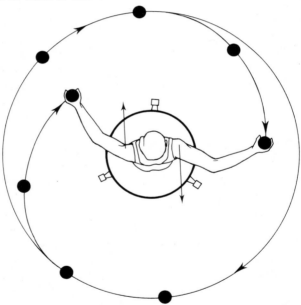

Fig. 21. Plan view of Experiment 1. The black circles show successive positions of small, hand-held masses in process of being brought in from their wide circular paths towards the central axis of the turntable as the latter rotates in the clockwise sense. The resulting 'couple' experienced at the shoulders is shown by the arrows. Centripetal forces are not shown.

However, if the platform itself has a rather high moment of inertia, the experimenter will notice that one shoulder is apparently being pulled forward, the other backward, as the action proceeds; and this leads to a deeper insight into the mechanics of the process—a transfer of most of the angular momentum of the more remote elements to those parts closer to the axis. The forces on the shoulders—eccentric as far as G and the vertical axis are concerned, and in a sense tending to speed up their rotation and that of the whole body and the turntable—constitute an angular impulse which is increasing the angular momentum of this part of the system; while its reaction, delivered to the remote masses via the arms, reduces the angular momentum of these components by an equal amount (Fig. 21).

Looked at in this way, the experiment provides an example of an exchange of angular momentum; and since the two remote bodies concerned in it are of small dimensions they behave very much as massive particles, having no 'local' angular momentum but only that of the remote form about the central axis. This they are able to exchange with the rest of the system when they are brought very close to this central axis.

Experiment 2. Exchange of angular momentum between extended bodies   For this turntable experiment we require apparatus capable of being given considerable 'local' angular momentum: something conveniently provided by a bicycle wheel fitted with axial handles, and with metal cable (or other suitable material) packed uniformly into the rim. The radius of gyration of this device about its axle is almost that of the rim itself; and this, combined with its substantial mass, ensures a big moment of inertia, I_G, so that even a moderate rate of rotation about the axle is associated with plenty of local angular momentum.

First, it is well to check that this form of angular momentum cannot be transferred from the rotating wheel to the rest of the system merely by moving its vertical axle closer to that of the turntable. For this purpose, the wheel is set rotating and then handed to the experimenter, its axle in the vertical direction, as he sits on the turntable. The only angular momentum now possessed by the whole system is thus the local form already given to the wheel and still confined to it. It will now be found that no matter where the vertical axle of the wheel is held—near to the central axis of the turntable or far from it—as long as this axle is kept vertical it is incapable of setting the rest of the system in motion. Unlike the small masses of Experiment 1, the wheel has, as yet, no angular momentum of the remote form (i.e. no angular momentum of G about the central axis), and it is only this form which can be transferred by this method.

To effect a transfer of local angular momentum, the experimenter can exert a 'braking' torque on the wheel by arresting its rotary motion with his free hand. By so doing he will find that he has set himself and everything else, including the wheel that he is holding, rotating around the central axis of the turntable in the same sense as that in which the wheel itself

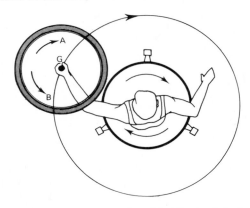

Fig. 22. Plan view of Experiments 2 and 3. The loaded wheel, already set rotating in the sense of the arrow A, is handed to the experimenter. Its local angular momentum can now be transferred wholly, or in part, to the whole system when the experimenter retards it.

Again, starting with everything at rest, the system can be set rotating about the central axis of the turntable when the wheel is set rotating about its axle in the sense of the arrow B. The rotary impulse delivered to it by the experimenter gives rise to the clockwise rotation of the system.

was rotating, the wheel's local angular momentum having been shared with the rest of the system.

It is now easy to transfer all the angular momentum of the system back to the wheel; for the wheel has only to be set rotating again in the same sense about its axle, with a rotary impulse delivered to it by the experimenter, in order to bring everything else to rest once more. In the absence of frictional effects the wheel will now be rotating as fast as it was initially, having received back all its former angular momentum without loss.

As with Experiment 1, it is well to examine the situation further. Although the *stationary* system is unaffected by changes in position of the rotating wheel (since the latter has no remote angular momentum to communicate to it), when everything is moving around the central axis, the wheel has angular momentum of both forms—'local', due to its own rotation about its own axle (something which stays constant when the axle moves), and 'remote', because its own centre of mass is remote from the central axis as it moves around it. It is the latter form which, as in Experiment 1, can be communicated to the rest of the system when the centre of mass of the wheel is brought nearer to the central axis of the turntable. The wheel, whether it is rotating or not, behaves just as do the small masses of the earlier experiment.

Experiment 3. Rotation with zero angular momentum   This is a simple modification of Experiment 2, but one having great importance in physical movement. Here the wheel is set rotating about its own vertical axle: not externally, as above, but by the experimenter himself as he holds it in this attitude on the stationary turntable. At no time does the system receive an external rotary impulse, so at no time does it possess angular momentum; yet the wheel has been set rotating by an internally-delivered rotary impulse, and this has given rise to an equal and opposite reaction that has set everything, including the centre of mass of the wheel, moving around the central axis in the opposite sense.

The experimenter thus finds himself carrying the wheel

around with him as he and the turntable rotate. The whole system—turntable, experimenter and wheel—is changing its attitude relative to fixed external objects without having any overall angular momentum, i.e. without having been made to do so by any external object. Unless frictional effects interfere, the act of bringing the wheel to rest will also bring everything else to rest simultaneously.

Experiment 4. The vectorial character of angular momentum
The convention whereby the direction of a vector, such as the angular momentum of a body rotating about an axis, is determined by the direction along the axis for which the body's rotation is seen to be clockwise, has already been stated. The turntable, sensitive only to changes of angular momentum in the vertical direction, will respond, therefore, only to changes in the vertical component of this vector.

Now it is clear that a wheel rotating about any horizontal axis has no angular momentum in the vertical direction; and that if the experimenter, sitting on the stationary turntable, sets the wheel rotating thus, the turntable will stay at rest. If, however, he forces the axle to turn into the vertical, he confers 'vertical' angular momentum on the wheel by this purely internal action, and so acquires, together with the rest of the system, equal angular momentum in the reverse sense. Returning the axle of the wheel to any horizontal direction *in any way at all*, will, in the absence of frictional effects, bring everything except the rotating wheel to a halt. This is a modification of Experiment 3.

Suppose that the wheel, already rotating about a *vertical* axis, is handed to the experimenter at rest on the stationary turntable, as in Experiment 2. By turning the axle into any horizontal direction he will eliminate all the wheel's local angular momentum in the vertical direction; but, as in the earlier experiment, this internal action can have no effect on the angular momentum possessed by the system as a whole, so everything is set rotating around the central axis in the same sense as that in which the wheel was rotating. The effect will be doubled if the wheel, instead of being given the limited turn described above, is forced to turn 'upside-down', when its initial angular momentum in the vertical direction will not merely be reduced to zero, but will be forcibly *reversed*.

Necessarily, whatever exchanges of angular momentum there are about horizontal axes in these experiments are exchanges with the Earth with which the turntable is in static contact.

Experiment 5. Rotary action and reaction about the body's longitudinal axis There is no specific way in which this investigation should be carried out: some aspects of it are more easily performed without a turntable.

Basically, the experimenter stands centrally on the platform with the feet spaced somewhat apart, to hold the body rigidly erect, and with the arms extended horizontally from the sides. When the arms alone are swung round horizontally in the same sense through as big an angle as possible, the rest of the body and the turntable rotate in the opposite sense. Clearly, the angular impulse given to the arms about the central axis at all times during their motion has reacted in the usual way on the rest of the body and on the turntable, to give these components angular momentum equal to that acquired by the arms but in the opposite sense.

This demonstration is similar to that of Experiment 3, but it differs from it in that whereas a wheel can make many revolutions before stopping, and can keep the experiment going for several seconds, the limited range of arm-movement available here soon results in both arms and all other components coming to rest simultaneously.

It will soon be noticed that, if the movement is performed in this way, the range of movement of the arms is limited by the contrary rotation of the shoulders; and that a more effective sweep of the arms is obtained when the shoulders and upper part of the trunk move with them. When this happens, the arm–shoulder–thorax combination will rotate in one sense about the body's longitudinal axis; the platform–leg–pelvis system in the other. It is also possible to split the system into three parts rather than two by driving the upper parts of the body in one sense, the hip region in the opposite one, and, by suitable twisting of the legs, to leave the platform at rest. The point to note is that the pattern of movement is decided, not by the relative moments of inertia of the relevant body-parts, or by any purely mechanical features, but by the type of muscular activity used to initiate it; although the angular distances through which it is possible to rotate these components about the longitudinal axis does depend on their relative moments of inertia about this axis.

Conservation of Angular Momentum

Mutual forces acting between component parts of a body, or of a system of bodies, have already been shown to have no effect on the *linear* momentum of the system as a whole; i.e. on the motion of its centre of mass (p. 54 and Fig. 13). This is because the individual linear impulses delivered between one part and another have, in accordance with the Third Law of Motion, the cancelling effect which equal and opposite forces always have over the same time-interval.

The above experiments strongly suggest the existence of a conservation law for *angular* momentum also: one based on the fact that angular impulses acting between parts of a body or between members of a system, one on another, have no effect on the angular momentum of the system as a whole about any axis, whether this axis is materially connected to the system or is an imaginary line in space. It is clear that this must be so, since, when mutual forces act between bodies, not only are the linear impulses equal and opposite (equal and opposite forces acting for identical times), but their lines of action are the same, and the torques they exert in opposite senses about any axis must therefore cancel. Only external rotary impulses can change the angular momentum of an isolated system of bodies about a given axis; just as, ideally, only an external one can interfere with the experiments just described: and, even then, must have a component about a vertical axis in order to do so.

As we have shown, mutual rotary action and reaction includes the exchange, or 'trading', of angular momentum between bodies in a system; but it must be realised that 'trading' can only take place when one of the bodies has a component of angular momentum in a direction about which the other is free to rotate. If this condition is not fulfilled, then the exchange is usually made with the Earth, and angular momentum *appears* to be lost.

One fact that is of great significance in human movement is the common location of the body's centre of mass and centre of gravity. Body-weight, having a resultant acting directly through G, can have no effect on the body's local angular momentum about any axis through G; so, under conditions of negligible air resistance—the only force which might possibly be eccentric—a body moving through the air preserves whatever local angular momentum it possesses: it cannot, by any means whatever, change its angular momentum about any axis through G; nor can it transfer its angular momentum about such an axis to any other one having a different direction in space. This makes even more remarkable the range and variety of movement which the human body is capable of exhibiting under 'airborne' conditions.

Problems

1. Show that all points on a uniformly-rotating disc are changing their directions with respect to each other at the same rate.

2. Consider the variation of linear velocity of points at different distances from the centre of a rotating disc.

A man attempts to run in a 'straight' line across the rotating base of a 'roundabout'. Explain the origin of the unusual radial and lateral forces he would experience. Relate these to the forces arising in Experiment 1, p. 81.

3. Show that the moment of inertia of a body is $m \cdot k^2$ about an axis for which its radius of gyration is k. If k is $1\frac{1}{2}$ ft for the transverse axis of an erect person, and his centre of mass, G, is $3\frac{1}{2}$ ft from his feet, at O, compare his moments of inertia about transverse axes through O and G.

4. A swimmer, toppling rigidly forward at the end of a diving-board, has local angular momentum about his transverse axis, and remote angular momentum about the board. Explain why the former stays constant and the latter increases once contact with the board has been lost. Hence show that the swimmer's angular velocity just before leaving the board is the same as its constant value afterwards, provided that no posture-change takes place.

5. Draw a diagram like Fig. 17, but with the line of action of the eccentric impulse (A′Z′) directed straight through O. Indicate the line along which G will travel, and show that the quantity $m \cdot v \cdot p$ is equal and opposite to $I_G \cdot \omega$, where I_G and ω refer to features of the body's rotation about G.

Show that, if the body strikes a rigid peg fixed at O, and bounces away from it with G moving directly away from O, it does so without rotating about G.

CHAPTER SEVEN

Angular Momentum in Human Movement

The importance of rotary motion

The principles discussed in the last chapter, and the experiments described, have been concerned with the application of Newton's Third Law to rotary motion, particularly to non-rigid bodies and systems of bodies. These applications are all relevant to the human body—one capable of adopting a wide variety of postures, and of controlling its attitude in the air. In particular, we note the two important facilities possessed by such a body: the power of altering its own mass-distribution, and therefore its moment of inertia about a given axis; and, even more significant, the possibility of making a redistribution of angular momentum among its component parts; so that, for example, limbs may be set moving around G with an angular velocity that they do not share with the trunk. These are powers which may be studied with apparatus on a turntable or with slow-motion films of human movement itself; their supreme virtue lies in the fundamental freedom the body has of continuing the *rotary* motion of its limbs indefinitely, as distinct from the limitation imposed on *linear* displacement by the finite length of those limbs.

What we are concerned with are the results of mutual actions between component parts of the body: effects which are far more spectacular when they are of a rotary nature than when they involve only linear displacement of one part with respect to the rest. Consider, for example, the effect on the human body in free fall when a strictly rectilinear movement of a limb is undertaken either directly towards G or directly away from it. This only displaces the rest of the body a very short distance on the other side of G, while this point, of course, remains unaffected. Linear displacement of the main part of the body is strictly limited, therefore, for it can only continue to increase as the extension of the responsible limb also increases: when this stops, everything stops. Furthermore, there is no possibility of restoring the original configuration of the body without at the same time eliminating the linear displacement.

The situation is quite different, however, in the case of rotary displacement. Suppose, for example, that a trampolinist making a high bounce without rotation were to start a 'skipping' action of the forearms—a rotary motion of these members about (for simplicity) the body's transverse axis through G. As with Experiment 3, p. 85, the rotary impulse required to start this motion must come from the rest of the body, so the angular momentum of the rotating forearms in one sense will be associated with equal and opposite angular momentum of the rest of the body; and there must occur, then, a slow rotation of the whole body about this transverse axis in the opposite sense: something that will continue as long as the arm-action continues.

Now, unlike the linear displacement discussed above, this arm-action can go on indefinitely—their 'angular' displacement has no limit in its sweeping out of complete revolutions; and so the slow somersaulting of the body about this axis will also go on until the arms cease their movement. Given sufficient time in the air, therefore, the body could make a number of complete somersaults without at any time possessing angular momentum about G: a range of movement having no parallel in the linear case.

Not only can rotation be initiated in the air, but, clearly, it can be controlled. An increase in the angular velocity of the trampolinist's body can be achieved by increasing that of the moving forearms or by enlarging the radius of their approximately circular path. Or he can do what the high-diver sometimes has to do, and reduce his moment of inertia about the transverse axis by adopting a temporary 'tuck' posture rather than an extended one. The resulting increase in angular velocity is reduced to its former value when the extended form is regained.

Application to recovery of balance We can now offer an alternative explanation of the 'balance-recovery' process discussed on pp. 50 and 51, where it was dealt with in terms of the derivation of a backward horizontal frictional force from the ground—a highly eccentric force as far as G is concerned —and associated not only with the forcing of G back over the feet, as desired, but with the accompanying accelerated rotation of most of the body around it (Fig. 12).

This line of argument can now be replaced by one which considers the trading of local angular momentum about the transverse axis through G for that of the remote form about the transverse axis through the feet at O, about which the unbalanced body is shown to be rotating in Fig. 12(a). The first thing to notice is that both normal ground-reaction, N, and friction, P, no longer enter the discussion since their lines of action both pass through the remote axis at O, and so have no turning-moment about it. Neither force can make any contribution to the body's angular momentum about O, the only force capable of doing this being the resultant weight of the body, acting vertically through G. It is this, of course, that is responsible for the accelerated departure of the rigidly-held body from the stable attitude when balance has already been lost, for this is the only force having a torque about O under these circumstances.

Another feature that has to be recognised in human movement is one which becomes apparent when we transfer our attention from the uniform rotary motion of symmetrical apparatus, such as wheels and turntables, to the altogether less predictable pattern of movement of the jointed parts of the body. Thus, in the posture-changes shown in Fig. 12 and elsewhere, it is unlikely that any parts of the body will move in simple circular paths around G; nor will they have the same angular velocities about it; nor will G itself be likely to move in a circular arc around our chosen transverse axis through O. Some loss of descriptive precision always arises, therefore, when body-movement is being analysed; but this does not invalidate the general argument nor give less confidence in the conclusions reached, for it must be remembered that angular momentum about a point or an axis depends on the sweeping-out of area about that axis, not on the shape of the area swept out.

The argument is now based on the principle illustrated in Experiment 3: i.e. the development of rotary motion around an axis without the acquisition of angular momentum. The practical conditions differ from those of the experiment in that the axes concerned (the transverse axes through G and O) are horizontal instead of vertical, and the one through O is far from the 'centre' of the non-rigid body. Of greater importance is the fact that in the off-balance attitude of Fig. 12(a) the body is not only not at rest, but its weight is putting it under the action of a rotary impulse, giving it increasing angular momentum about O. It is this angular momentum that has to be reduced to zero and then reversed in order to bring G back vertically above O, as in (c), having passed through position (b).

Referring, first, to (a), we see that the body is moving like the rigid one discussed on p. 73, and has angular momentum of both the local and remote forms in the anticlockwise sense about O: local, since the whole body is rotating (i.e. changing its attitude) about G; remote, because G is itself revolving around O. If, to emphasise the principle involved, we temporarily ignore the torque which body-weight, W, has about O, then we have the body rotating 'forwards' at a *constant* rate; i.e. with no build-up of angular momentum as it topples. All that is needed to stop this movement and to bring G to rest is for the body to break at the hip so that most of it is made to rotate much faster about G, as in (b); for this is a purely internal action capable of removing all the body's remote angular momentum in favour of an equal increase of the local form: just one of the phenomena shown in turntable experiments.

However, it is not enough merely to bring G to rest: local angular momentum has to be further increased so that G's motion is not only stopped but reversed; and the turning-moment, hitherto disregarded, which body-weight has about O must now be considered. We have seen that without this turning-moment there will need to be only a steady, uniform posture-change about G to confine all the necessary angular momentum to the local form, and so bring about recovery; but, as seen in (a), the weight is delivering to the body a rotary impulse which is increasing as time goes on. It follows that the body is being given an increasing amount of angular momentum in the unwanted sense: something that can only be confined, or stored, as local angular momentum if the rotary motion about G is an *accelerated* one; and it is this fact which makes recovery so much more difficult to achieve.

The accelerated breaking at the hip has to go on until the vertical line of action of W has come back to the right of O, as in (b); for its inevitable retardation to rest, and then reversal towards the upright stance of (c), will move it forward again to its required position over the feet. It should be noticed that the posture-change involves nearly all the body, but the supporting leg is antagonistic to it in that its rotation about G is in the opposite sense to the one required.

The mechanics of the step forward from rest

We now come to a body-movement designed to have an effect opposite to the balance-recovery one discussed above; for here, the act of moving one leg forward is meant deliberately to throw the body *off-balance* in this direction. At first sight there might seem to be no need for further investigation, for obviously an accelerated posture-change in the clockwise direction, i.e. opposite to the one shown in Fig. 12, would provide the necessary local angular momentum about G; and the freely-moving leg is doing this (Fig. 23). However, this cannot be taken for granted: we have just noted that even the balance-recovery technique causes some part of the body—the supporting leg—to move around G in the wrong sense; and it is quite possible that the forward-stepping rotation of a leg about G may have its effect either cancelled or reversed by a simultaneous forward lean of the rest of the body—an unwanted anticlockwise movement around G.

Before investigating this, we must realise that we are not dealing with the usual practical method of moving forward *from rest*. In the absence of any other conscious effort, the striding forward with one leg does not normally start until the body has been made to lose balance forwards, and is already moving: initial acceleration is the result of the turning-moment of body-weight about O, just as it is in Fig. 12(a). As before, we disregard this as a separate effect, and consider whether the rigidly-held body, balanced and at rest, will sway back as in Fig. 23, or will be made to lean forward, when one leg is raised as shown. If the former, then virtually all the body will have the desired clockwise movement around G, and G will move forward (anticlockwise) with respect to the feet at O; if the latter, then the matter will still be in abeyance. The experimenter may like to make a practical test, before coming to a theoretical conclusion, as to whether the *immediate* effect of the leg-movement is as shown (very much exaggerated) in Fig. 23, or not; but it is not easy to decide, even by watching one's profile in a large mirror.

The theoretical argument is somewhat different from the one used hitherto. Instead of considering the equality of the local and remote forms of angular momentum as they arise in opposite senses about G and O respectively, we use Fig. 23 to see how the small individual elements of the leg move with respect to O; i.e. the sense in which they sweep out their own individual areas around O: the sense in which they possess angular momentum about O. Now, even if there is a sway-back of the rest of the body, it is clear that nearly all parts of the active leg have angular momentum in the anticlockwise sense about O, the only notable exception being the foot-and-ankle. As we have claimed so often, the purely internal effort of moving the leg cannot give the body any overall angular momentum about O; nor can the ground-reaction, acting *through* O; so it follows that the rest of the body must do what

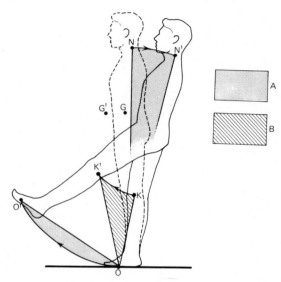

Fig. 23. Areas swept out in different senses about O by elements of the foot, the knee, and the neck, as one leg is raised. If they predominate in the *anticlockwise* sense for all parts of the leg, then the trunk will sway back about O, and nearly all the body will rotate 'backwards' (clockwise) with respect to G, which will move forward to G'. Change in trunk-attitude has been greatly exaggerated for clarity.

The shading of the rectangles A and B indicates the sweeping-out of area in the clockwise and anticlockwise senses, respectively.

Fig. 23 suggests, and rotate backwards in the clockwise sense, carrying with it angular momentum precisely equal to that developed in the leg. Necessarily, the resulting forward movement of G immediately gives body-weight a 'forwardly'-sensed turning-moment—the one disregarded above—which gives further forward acceleration to G.

Control of attitude by 'cyclic' limb-movement

The 'skipping' action of the forearms of a trampolinist has already been used (p. 92) as a simple illustration of the possibility of using a continuous form of rotary movement to promote a change of attitude in the air; and we know why such movement is equally effective—like that of the rotating wheel in turntable experiments—when it takes place about an axis far from one through G, instead of coincident with it. The effects generally observed, unlike those of the turntable, are most commonly found in the control of body-rotation in the vertical, sagittal plane when the body is freely moving in the air: they usually involve full circular sweeping of the arms about the shoulders, together with exaggerated aerial-striding of the legs. The well-known 'hitch-kick' technique in long-jumping is an excellent illustration of the principle, and will be used as the basis of this discussion.

We have noted earlier that the jointed human or animal body lacks the clear-cut simplicity of motion of the loaded wheel and turntable, so its behaviour sometimes has to be studied in terms of components of its motion resolved in directions which happen to be convenient. These are often the body's own principal axes, or, as in this case, axes very close to them (Fig. 24). Resolved components of a body's motion are easily visualised if we imagine ourselves to be stationed at some distant point so that we see the motion in a direction at right-angles to the plane in which it is being studied: in Fig. 24(a) the point of observation is on the horizontal axis through G, at right-angles to the plane of the body's motion. All displacements, velocities, accelerations, and every other vector concerned in the motion will now be projected for us in this plane; and all measurements we may make on these quantities will be made on their components resolved in it. All movement will be seen as if the whole body had been compressed, or laminated, in this direction, at right-angles to our line of sight; and no movement *along* this line will be apparent to us.

The first thing to remember is that, in the absence of air

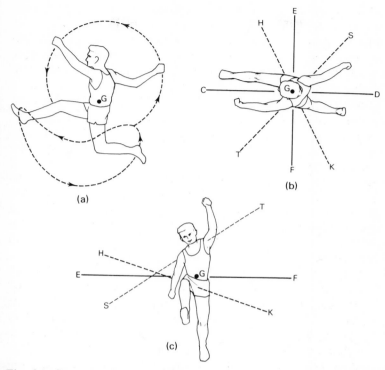

Fig. 24. Control of trunk-attitude in long-jumping: simultaneous views projected in perpendicular directions. Where relevant, principal axes through G are shown in the notation of Fig. 19.

resistance, there is nothing the subject can do to modify in any way the path of his centre of mass, G, in the air: this has already been determined by events leading to the take-off. We know, also, that all parts of the body are, like everything in free fall, mutually weightless since they experience the same downward acceleration, g, common to all such bodies. These facts make the axis shown in Fig. 24 an ideal one to take the place of the central axis of a turntable, with the continuous cyclic motion of the arms and legs to simulate the behaviour of the loaded rotating wheel.

If this is accepted, then we can keep Experiment 2 (p. 83)

in mind to understand how the long-jumper, having left the board with angular momentum in the anticlockwise sense (Fig. 24), is able to do what a rigid body cannot do, and start the limbs rotating in the same sense and at such a rate as *to take to themselves* this angular momentum. As with the turn-table experiment, rotation of the trunk then ceases, and it maintains its nearly erect attitude until the landing is about to be made.

In a similar way, even if the jump has been made without this unwanted angular momentum, rapid correction of trunk-attitude can be made by the same technique, the limb-rotation being started in the opposite sense to that of the correction required, and stopped when the desired attitude has been restored. This will be recognised as an application of Experiment 3 (p. 85).

Motion in other planes   Control of attitude in the air has so far been dealt with in relation to the vertical plane of a long-jumper's motion—the one of greatest significance in this and in numerous other activities.

In the horizontal plane the projection would appear some-what as in the plan view of Fig. 24(b); and from this it is clear why the leg-movement ought to accompany that of the arms, and why it must be out of phase with it. In this diagram, for example, the line of the hips, HK, is about to move in the clockwise sense as the right and left legs start to exchange their forward and backward extensions, respectively; while the line of the shoulders, ST, is due to start moving in the opposite sense with the arms. These pendulum-like move-ments of the limbs on opposite sides of the body—as they are seen to be in this projection—constitute angular momenta about the vertical axis through G; and these must at all times be equal and opposite if there is to be no undue disturbance of the other axes, CD and EF. The rotary impulse that sets the arm–shoulder system rotating in one sense must have its reaction in the equal one that sets the leg–hip system rotating in the opposite sense, the trunk remaining relatively un-disturbed.

The situation as seen projected in the frontal plane (Fig. 24(c)), shows a similar anti-phase movement between HK and ST. Any difference that may exist between the angular momenta carried by the two separate movements on opposite sides of the sagittal plane will be taken up by lateral swaying of the head and trunk—a motion that is not disturbing to the jumper.

One important point to bear in mind is that if some form of limb-movement is used to prevent a progressive change of attitude in the air by 'absorbing' the angular momentum associated with it, then this movement (or one of equal effectiveness) must be continued as long as the body is free from the ground. If this is not done, the body will again start to change its attitude in the same way as before. If, however, there is a 'statically' incorrect attitude to be adjusted—one that is not changing with time—then this can be done by a rotary limb-movement which lasts only as long as it takes to make the correction.

Elementary treatment of the forward pike in the air

In the context of long-jumping, the cyclic motion of the limbs is seen as a means of maintaining correct trunk-attitude when the jumper is embarrassed by angular momentum in the 'forward' sense. The final forward piking of the body prior to landing, however, is necessarily undertaken without this stabilising influence, and, when forward body-rotation re-asserts itself, there is a tendency for the legs to come up very slowly, if at all.

Yet, even if the body of the jumper is free from angular momentum generated on take-off, it is still possible for the piking action itself to be technically faulty and to produce a premature dropping of the legs. In studying this we have to realise that the forward rotation of the upper part of the body about the hips, and the simultaneous forward and upward movement of the legs, are mechanically more complex than, for example, the closing of the jaws of a crocodile or of a pair of wire-cutters. These have fixed axes about which to rotate, whereas the transverse line through the hips of the 'airborne' athlete—the axis about which piking normally takes place—is free to move upwards or downwards as the trunk and legs approach each other; and its direction of motion depends on more factors than one. Clearly, this direction should be downward with respect to the body's centre of mass if the feet are to be kept up.

In this simple treatment we first suppose the body to be of a much simpler form than it really is: we let its upper part (Fig. 25) be represented by a point mass located at its own centre of mass, G_1, and the legs similarly by their mass at G_2. Then before the motion starts, and at all times after this, G_1 and G_2 will be disposed on opposite sides of G, the centre of mass of the whole body; and they will always be placed so that all three points lie on a straight line. Now 'point' masses can have no local angular momentum; and because the body itself is assumed to have none about its own centre of mass, G, it follows that G_1 and G_2 cannot move in any way except towards G or away from it. This means that the line joining them remains in the same direction as they move simultaneously towards G during the piking movement. We note, with reference to Fig. 25(a), that the path of G in the air is unaffected by the internal impulse causing the posture-change: only the lines joining G_1 and G_2 to the hip-joint, H, will be affected; and examination of the sequence of line-diagrams will show that if G_1H is longer than G_2H, then the hips will tend to drop as the legs rise. Using this much simplified model, therefore, we are given the suggestion that correct technique requires maximum distance between the centre of mass of the upper body and the hip-joint; and this will be achieved by keeping the back as flat as possible with the arms extended in line with it. Effective and ineffective forms are shown in (b) and (c), respectively.

The more accurate treatment, in which the masses of the upper part of the body and the legs are regarded as distributed masses rather than points, is one that takes into account the finite moments of inertia which these masses have about the transverse axes through each of their centres of mass (just as a wheel has about its axle). This means that not only is each centre of mass brought nearer to G as the piking action proceeds, but each part has to be set rotating about its own

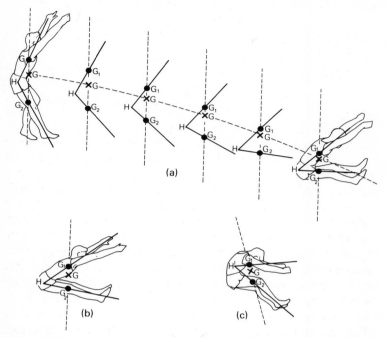

(a)

(b)   (c)

Fig. 25. The lifting of the legs in long-jumping is helped, as in (a), when the centre of mass of the upper body is further from the hips than that of the legs. The posture adopted in (b) is effective; but a piking-action about an axis above the hips, as at H' in (c), fails to raise the legs. The technique will be even more effective if the legs are bent at the knees as the action proceeds towards the final position shown in (b).

centre of mass, and so has its own local angular momentum brought into the argument. The result of this is complicated, and beyond the scope of this book; but it can be shown to make no essential difference to the conclusion already reached.

The landing from a vault or a jump

Here we shall assume that, in general, a vault may be recognised by the following basic features:

(1) a run-up to a take-off point from the ground or from some form of resilient apparatus;

(2) a flight-path in the air which at some stage is assisted by hand-contact with some fixed, elevated structure such as a vaulting-horse;

(3) a final, controlled landing, usually finishing in an erect, balanced attitude.

In pole-vaulting, the 'elevated structure' is the planted pole itself, and no particular landing attitude is specified. This athletic event will be treated separately. The term 'jump' implies no contact with apparatus between take-off and landing, and the latter may not necessarily be a controlled one.

The elements of a controlled landing are dealt with in connection with Fig. 26, where ground-contact has been made at O, with G in the position shown. G is assumed to be moving with velocity v in the direction indicated by the appropriate arrow. The body evidently lands with some remote angular momentum about the transverse axis through O, its magnitude being $m \cdot v \cdot p$ in the clockwise sense, as in Fig. 17 (p. 71), but it is assumed, at this stage, that there is no angular momentum of the local form; i.e. there is no rotation of the body about the transverse axis through G. Now it has already been stressed that a force such as ground-reaction, R, no matter in what direction it may act, can have no effect at all on the total angular momentum of the body about O, since, acting at O, it has no turning-moment about it. What it can do, however, is to cause one form to be 'traded' for the other, and this it does when it is an eccentric force with respect to G, as it is shown to be in Fig. 26(a).

Let us suppose, temporarily, that R is the only external force acting on the vaulter as he lands, and that he is required to finish his vault in a final erect attitude with G at rest vertically above O, at G' (Fig. 26(a)). To bring G to rest in this way, R must have a backward inclination, and it must be big enough to provide the necessary checking impulse in the short time available: it must get rid of all the body's *linear* momentum, $m . v$; and its direction must therefore be opposite to that of G's velocity. In bringing G to rest, then, the force R has eliminated all the body's remote angular momentum about O—the product $m . v . p$ is now zero—but, as an eccentric force, R has simultaneously developed equal angular

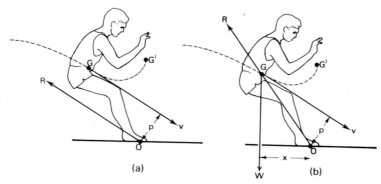

Fig. 26. In (a), *the weight of the gymnast is entirely ignored.* Ground-reaction, R, can cancel his linear momentum but not his total angular momentum about O.

In (b), R can still help to cancel linear momentum, and the turning-moment of W about O can remove angular momentum if the landing is skilfully made.

In all satisfactory landings, G must come to rest at some point, G', vertically above O.

momentum of the *local* form about G; so, although the body is poised with G at rest where it should be, some part of it (e.g. the arms) will need to be kept constantly rotating in the clockwise sense, to 'absorb' this local angular momentum. Furthermore, under the conditions assumed here, there is no way whereby the body can be rid of this embarrassing angular momentum, for any attempt to reduce it will immediately cause some retransformation back to the remote form, with the inevitable forward movement of G once again.

Fig. 26(b) shows how important the turning-moment,

$W . x$, is in leading to a good landing. This turning-moment about O operates until G is vertically above O, and the rotary impulse due to it depends on how its lever-arm, x, is allowed to diminish with time. If this is skilfully controlled, then the body's remote angular momentum about O can be cancelled exactly by this rotary impulse, and all that ground-reaction is required to do is to provide a *direct* force through G which has horizontal and vertical components capable, with W, of bring G to rest vertically above O. G will then be at rest, and the body will not be rotating about it.

Other possible landing patterns   Diagrams like those of Fig. 26 can only give an overall picture of what should happen under the given conditions. The magnitude and direction of the force R, for example, do not necessarily stay constant while G is brought to rest in the perfect landing of Fig. 26(b): all that is required for this is that the impulse delivered by R shall have the necessary magnitude and the *resultant* direction: it need not pass through G all the time. We should note that skilful control of R at all times during the short period of the landing is the only way the vaulter has of making a reasonably satisfactory landing—he cannot *directly* alter any other factor.

Suppose, for example, that in (b) the velocity is excessive or the perpendicular distance, p, is too long. Normally, with more remote angular momentum about O than the contrary rotary impulse of the weight can successfully cancel, the body will topple forwards over the feet, and a step forwards will need to be taken. However, if G's excessive velocity can be reduced *early*, so that the lever-arm x is kept as long as possible over as long a time-period as possible, then the cancellation may be possible; and this can be done by changing the direction of R so that instead of being a direct force through G it is more backwardly-inclined beneath it, as in (a), and has a bigger horizontal checking component against the motion of G. Unfortunately, to redirect R in this manner, there will need to be an accelerated clockwise rotation of the upper part of the body about G; and this will have to be continued for as long as R has to have this new, eccentric direction. Later, with G nearly at rest, the upper-body rotation can be retarded and reversed (R's line of action going up over G), and recovery made complete. This is not a perfect landing because the 'remedial' effect of local body-rotation has had to be invoked, but it illustrates a successful method of control that is often seen in gymnastics.

A further common landing-condition is one in which touch-down occurs with too short a horizontal distance between G and the feet, and the necessary horizontal checking component cannot be obtained, as it was above, by a redirection of R because too much local angular momentum would be generated about G. Under these circumstances an alternative solution is sometimes tried: a drive downwards against the ground vigorous enough to produce a big increase in R without altering its direction through G. If R is made big enough, then its backward horizontal component may still be able to bring G to rest over the feet at O. Under these circumstances, however, the *vertical* component may now be so big as to drive the body clear of the ground so that, although all rotary effects have been cancelled successfully, and the erect attitude achieved, a small vertical jump is the final action: not a perfect vault, but one seen fairly frequently.

In all landing patterns considered above, the body has approached touch-down with no local angular momentum about the transverse axis through G. Clearly, the corrective measures described will be more difficult to apply if there is some such momentum in the 'forward' rotary sense: on the other hand there are instances, as in long-jumping, where the feet have to land as far forward of G as possible, and where angular momentum in this sense helps to prevent falling-back into the pit. In this case, G is encouraged to maintain its horizontal progress rather than to be checked, so the severely-piked body must 'give' at the knees and hips to make the initial value of R as small as possible, and the backward inclination of this force must be reduced. Evidently, a favourable direction of R will be associated with an accelerated backward rotation of the upper body as the feet meet the sand,

and this should continue until G has successfully passed over them.

The 'association' of force with acceleration   It is common practice, and very natural and convenient, to regard force as the cause of a body's acceleration, and a turning-moment as responsible for its *rotary* acceleration. Nevertheless, in much of what has been described above an external force has only arisen *as a result of* the accelerated posture-change which the body has undergone, and has been determined, in magnitude and direction, by the nature of that posture-change. Similarly, the rotary acceleration of part of the body about G is actually used to develop exactly the right eccentric force that would otherwise be regarded as causing it. Thus, although the relationship between force and acceleration is more obviously regarded as a causal one, it might be more correctly interpreted as an intimate 'association'. Force is *associated with* the acceleration of a massive body: it does not necessarily cause it.

Earlier phases of a vault

The final stage of a vault has been discussed first because it shows what has to be done in the earlier phases to make a controlled landing possible. The purpose of the take-off and of the effort made during contact with apparatus is not, of course, the reduction of all linear and rotary motion to zero, but in each case it is the projection of G in the correct parabolic path in the air, together with the provision of local angular momentum capable of rotating the body into a suitable attitude for contact, first with the apparatus, and then with the ground.

Take-off   No doubt there are many different force-patterns capable of being used effectively during a running take-off, and it is only necessary here to point out some basic features common to such actions. For a vault like that of Fig. 27, the body has to leave the ground (or whatever take-off apparatus may be used) with local angular momentum in the forward sense, and this will be preserved until hand-contact with the obstacle is made. During most of the take-off time, therefore, the line of action of R will pass up behind G, developing the big clockwise rotary impulse required; this suggests a backward inclination of R, causing partial checking of the body's forward speed. The way the direction of R changes as G rides

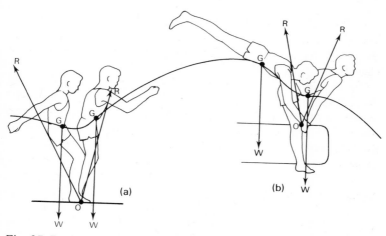

Fig. 27. During the vaulting take-off, shown in (a), one essential function of R is to give the body enough local angular momentum to rotate it into the attitude of (b), where contact with the apparatus has been made. Here it is the function of W to remove nearly all the 'forward' angular momentum about O so that a suitable landing may be made. R controls the rotary impulse due to W by controlling the way in which its lever-arm changes with respect to O.

over the feet depends on the accelerated changes of posture made by the body at this time. Fig. 27 shows the force directed nearly through G as the final extension of the body is made, so that minor correction of its local angular momentum, in one sense or the other, may be made before ground-contact is lost. Body-weight, which changes the sense of its turning-moment about O from anticlockwise to clockwise, has no great effect on the body's angular momentum because here it is a comparatively small force compared with R; its line of action is never far from the contact-point, O; and it acts on both sides of it in turn.

Contact with apparatus   Figs 27(a) and (b) are useful in reminding us of the relative rôles played by body-weight, W, always acting through G; and 'ground' reaction, R, always acting through O. W is the force that can change the body's remote angular momentum about O (but it cannot change the *local* form): R cannot change the body's *total* angular momentum (i.e. remote form plus local form) about O; all it can do is to increase one form of angular momentum at the expense of the other.

Now, whereas in (a) the eccentric character of R is meant to rotate the body 'forwards' about G so that it contacts the apparatus in the right attitude, in (b) its function is to eliminate all or most of this local angular momentum; otherwise, as we have seen earlier, a good landing may be impossible. If it is objected, after examination of (b), that W is itself changing body-attitude in the anticlockwise sense about O, as it appears to be doing, then it must be pointed out that body-attitude would not change in the slightest if the force R were absent. It is only the motion of G around O (i.e. its own angular momentum about O, expressed in the form $m . v . p$, as it has been expressed earlier) that W is changing by virtue of its torque about O.

Control of the torque due to body-weight   What has just been said does not in any way minimise the importance of R as a means of controlling the *linear* motion of G, and, as in the final landing, increasing or decreasing the time for which W is allowed to exert its torque about O. As in landing, the magnitude and direction of R are the only factors under the control of the vaulter, and he controls them by making the appropriate posture-changes at the right time. In (b), for example, the initial direction of R is one having a checking effect on G's forward velocity, so the time for which W can exert its anticlockwise torque about O will be increased. This may well be desirable, for while R transforms most of the body's clock-wise angular momentum from the local form to the remote one, W is given a better chance of removing enough of this to keep the term $m \cdot v \cdot p$ down to a manageable value.

Again, as G moves forward over O, R's direction changes with it, so that a minor posture-change gives R the final small controlling effect before contact with the apparatus is lost. It should be noted that, in hand-contact, R is more nearly of the same order as W, so W has a bigger relative effect in this operation than it does in take-off.

Pole Vaulting

Fundamental considerations

A little thought will show that the clearance of a high bar with the aid of a pole can be discussed in much the same terms as those used above for the clearance of a piece of gymnastic apparatus. In its initial action after the take-off from the ground, a flexible pole performs the same mechanical function as a beat-board or trampette; in the nearly vertical position, it also has some of the supporting properties of the vaulting obstacle itself. Needless to say, landing attitude is unspecified.

It is fruitful to compare the mechanical features of the pole-assisted vault with those of the gymnastic one just discussed; we first note that in the latter case there is no possibility of success if the 'adverse' torque due to body-weight (Fig. 27(b)) is given time to remove all the body's angular momentum about O before G clears it. In particular, this can be caused if G is checked excessively in its forward progress by an abnormally big backward thrust due to the reaction, R. So it is with pole-vaulting: there is no possibility of the vaulter's centre of mass, G, passing vertically over the grounded end of the pole, O (Fig. 28), unless he has retained at least part of the remote angular momentum he gained about this point during his run-up and take-off. His fundamental problem, then, is to acquire a great deal of angular momentum about O, and to preserve as much of it as possible during the 'pole-borne' phase of the vault, so that, at its high-point, G is in the neighbourhood of the bar and still moving forward at a convenient speed: it must not fall back.

Acquisition of angular momentum about O Consider how angular momentum about the lower end of the planted pole arises. During the run-up to the take-off point, G moves with

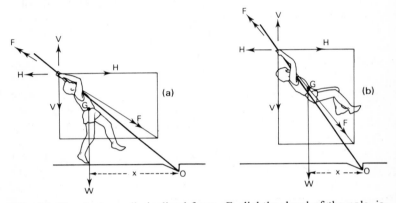

Fig. 28. The downwardly-inclined force, F, slightly ahead of the pole, is the force exerted *by* the vaulter *on* the pole. Its H and V components, and the reactions of all these forces on the body of the vaulter, are also shown. The big lever-arm, x, which determines the adverse torque due to W, shows why H must be made as small as possible in the early stages of the vault. Ground reaction, R, is not shown: its direction depends on what the pole is doing at the time, although it must always be closely aligned with the pole.

increasing speed in a somewhat undulating but nearly horizontal path which, if continued, would take it over the point O and about 4 ft above it: this is the distance we have called p in the expression m . v . p, and we see that at all times when G is moving horizontally during the run-up, the body's remote angular momentum about O is limited by this rather small distance-factor. The motion of G as a massive particle is not really the only form of angular momentum possessed by the running vaulter, for the striding action of the legs—moving in closed paths with respect to G—itself contributes a small amount (of the local form); this becomes available to the body when the striding-action stops. It should be remembered that a body can have angular momentum about a remote point or axis without being connected to it in any way. (See Figs 16 and 17, pp. 68 and 71.)

A fast run-up is not the only action the vaulter takes to gain maximum angular momentum about O: the drive from the ground as he transfers himself to the pole at take-off makes a further important contribution to it, for if this upwardly-inclined thrust changes G's direction of motion without reducing its speed, then there will be a considerable increase in the perpendicular distance, p, referred to above, and a corresponding increase in the body's angular momentum about O. If G can be speeded up, so much the better, for there must be no undue delay in getting G vertically above O, as will soon be apparent.

Preservation of angular momentum about O   Reference to Fig. 28(a) will show that the situation of the pole-vaulter during his 'pole-borne' phase is basically the same as that of the gymnast in the early stages of hand-contact with his apparatus. As we have emphasised, the reaction, R, of the ground or of the apparatus cannot alter the term $m \cdot v \cdot p$, which expresses the body's remote angular momentum about O, since in both cases it acts *through* O. On the other hand, it does develop a backwardly-inclined *linear* impulse on the body which retards the forward progress of G.

In the gymnastic vault this may well be desirable to some extent; and, as we have seen, variation of R can even be used to control the performance of the vault; but where maximum height is required, as in pole-vaulting, R's adverse checking effect must be minimised, especially in the early stages where its backward inclination is pronounced. This is of vital importance here, where the horizontal lever-arm, x, between O and the line of action of body-weight is initially so great (Fig. 28); for, as soon as the vaulter loses the support of the ground, the enormous torque which his weight has about O starts to build up a rotary impulse in the sense contrary to that of the body's angular momentum (anticlockwise in Fig. 28); this continuously reduces its fund of angular momentum as time goes on. It is easy, then, to see why G's horizontal position must be allowed to change quickly when G is still far from the vertical line through O; and success in this depends on how fast G is already moving in this direction after take-off, and what skill the vaulter can use in co-ordinating body-posture and the properties of his pole in order to reduce the magnitude of R.

The behaviour of a light pole

It is fortunate that the pole carried under competitive conditions is very light—little more than 4 lb wt—so that, without much error, its weight may be neglected. Its length—about 16 ft—gives it, however, a moment of inertia about one end which could be as much as 15 per cent of that of the standing vaulter about an axis through his feet. It can, however, be shown that at no time during a normal vault is the angular momentum of the pole about its lower end, O, more than about 3 per cent of that of the vaulter about the same point. When the argument is based on angular momentum, there is little error if, like its mass, the angular momentum of the pole is disregarded.

   These small values show us that the planted pole will tend rapidly to align itself close to the line of action of any single big force applied to it near its upper end; only when such alignment is achieved will the pole be in a condition of equilibrium. This may be shown easily with a more manageable device, e.g. a javelin with one end resting on a non-slip surface and a length of cord attached near the other, to support it in any inclined attitude. When the cord is put under considerable tension, the javelin will rapidly change its attitude until its length coincides with the cord, no matter where the free end of the latter is held. If the tension were great enough to bend the javelin, then equilibrium would be achieved with the force always directed towards the lower end: i.e. along the *chord* of the bent portion. If, therefore, a vaulter takes a very close hand-grip on his pole and exerts what is effectively a single force on it, the pole proceeds continuously to carry him up towards its vertical position if at all times this force is directed very slightly ahead of the pole or its chord. If it is not, the motion of the pole will soon be reversed.

It is profitable to analyse this situation by resolving the force exerted *by* the vaulter *on* the pole into its horizontal and vertical components, as in Fig. 28. The importance of this is that it shows the force, F, to have the required direction (very slightly ahead of the pole or its chord) only when its horizontal and vertical components, H and V, have the required ratio. In the early stages, H is considerably greater than V, as in (a); but as the elevation of the pole changes, it becomes less, as in (b). The early situation is evidently most unfavourable to the vaulter, for although the pole is exerting on him a suitable supporting force equal and opposite to V, it is retarding his forward progress with a much bigger backward reaction equal to H, and this at a time when, as we have shown above, checking of forward progress has to be minimised. In the circumstances assumed here, the only thing the vaulter can do to improve the situation is to reduce the upward force, V, that he needs to support him; for, if this can be done, then both F and H will be correspondingly reduced. Now V is greater than body-weight because G is not only *moving* upwards: it is *accelerating* in that direction as the body swings around the hands. Clearly, the body's local angular momentum about G is being increased in the anticlockwise sense by the eccentric force, F, acting on the hands; but the increasing rate of rotation can be reduced by the adoption of an extended body-posture, with its attendant big moment of inertia about G. This is the function of the 'hang' style at this stage of the vault: to reduce G's upward acceleration.

Now, if the pole bends somewhat under the force applied to it, not only can G's upward acceleration be reduced: it may even be reversed, and V will then become *less* than body-weight. This is possible, even if the pole remains rigid, if G's path immediately after take-off is allowed to level off somewhat; for, while this is happening, body-weight is not being entirely supported, and all the forces in Fig. 28(a) become temporarily less. A bending pole accentuates the effect and keeps G's trajectory low for a longer period. This is the great virtue of a really flexible pole, for its bending is capable of making a big difference to the form of the $m . v . p$ term in the early stages of the vault: a low and comparatively level path of G is associated with a small value of p and a correspondingly large value of v—the effect urgently wanted.

Control of a flexible pole Experiments with a fibre-glass pole—which is really a thin-walled tube—show that it is almost perfectly elastic: i.e. the force required to bend it at a reasonable rate is not much greater than the force it is capable of exerting when allowed to recover at the same rate. This means that under vaulting conditions it will regain its straightness in the later stages of a vault almost as quickly as a theoretically perfect pole, and so it will be possible to allow G to move further forward in its fairly low path before being moved rapidly up towards the bar. Notice that the longer this is delayed, the shorter will be the perpendicular distance, p, and the greater the value of G's velocity at all subsequent stages of the vault (Fig. 29); much is therefore lost in the use of a rigid or less-perfectly elastic pole.

The property of the pole which determines the vaulter's ability to control its bending is that which relates its stability under a given compressive force, F, to the distance from the lower end, O, at which the force is applied. With a pole having the characteristics mentioned, it is found that when a big enough force is applied at the point A, or higher (Fig. 29(a)), the pole bends under it; but as the distance from O is decreased, a point B is reached below which this same applied force is unable to start the bending. Applied at B, then, or above it, a compressive force of this magnitude will be able to bend the pole: applied below B, it will be unable to do so—a bigger force will be needed. Once started, the bending of one of these poles will continue to breaking-point unless the applied force is reduced or its point of application is brought nearer to O.

Consider the limitation imposed by the use of the close hand-grip of Fig. 28. If, in (a), the pole is just capable of sustaining the force, *F*, without bending, then *no greater force than F* will be available in the 'swing-up' stage shown in (b), when *G* needs to be given maximum acceleration in the vertical direction. Clearly, this acceleration is limited by the

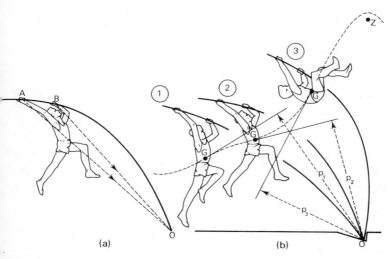

Fig. 29. In (a), a force applied to the pole at A has a better chance of bending it than an equal one applied at B. Control of the pole depends on the distribution of force between A and B.

In (b), the changing direction of *G*'s path is shown by the broken line leading over the bar, Z. The diminishing values of the factor *p* in the term *m . v . p* are shown for three positions of the vaulter.

vertical component, *V*, which cannot be made bigger without the pole bending, and continuing to bend unless relieved. Under these circumstances, control of pole-bending is only achieved by limitation of performance.

However, if a *wide* hand-grip is taken at the chosen part of the pole, using the points A and B (Fig. 29(a)), the nearly longitudinal force applied to it by the vaulter is shared between his hands and so made to act at two different distances from O instead of one: distances at which the pole responds in different ways to forces of the order of those developed in vaulting. The initial bending of the pole would be encouraged by applying most of the force at A; and, later on, when rapid recovery is required (position 3, Fig. 29(b)), the lower point, B, can be made the source of most of it.

Transformation of angular momentum between the local and remote forms

We have seen that in the 'hang' phase of a vault in which a close hand-grip is used, while G is well below the line of action of the backwardly-inclined force, F, acting on the vaulter (Fig. 28(a)), the resulting increase in the body's local angular momentum shows itself in a slow, accelerated forward swing of the extended body in the anticlockwise sense. This movement, designed basically to delay G's upward acceleration, is also a 'trading' of local angular momentum in this sense for equal remote angular momentum in the opposite sense about O: the sense in which it is urgently required at this early stage.

Later on, when G moves into the line of action of F, and then over it towards the body's inverted attitude, some local angular momentum will be lost: a process that simultaneously removes an equal amount, previously gained, of the remote form about O. Nevertheless, this retransformation takes place when the pole is nearly vertical, and slower progress of G towards the bar can be tolerated.

It is evident that the use of a wide hand-spacing on the pole would permit the use of *lateral* components of force exerted on the pole by each hand in opposite directions—forces at right-angles to F, and having the effect of delaying further the forward swing of G in the early stages and of speeding it up when required to do so later. This measure of control also tends to bend the pole early, to control its bending, and to assist its more rapid straightening during the 'pull-up and push' phase.

Problems

1. Carry out the experiment suggested in connection with Fig. 23: (a) with the trunk held in rigid alignment with the straight supporting leg; (b) with the supporting leg held less rigidly at the knee so that it can bend, if necessary, when the forward step is taken by the other one; (c) with the forward-moving foot travelling in a straight line from the ground instead of sweeping out the area shown in Fig. 23.
Account for the increased sway-back of the trunk under conditions (b) and (c).

2. The argument of Fig. 23 is based on the fact that nearly all the body has acquired local angular momentum in the 'backward' (clockwise) sense about G, and that an equal amount of the remote form has been generated, therefore, in the opposite sense, so that G moves forward. Using Fig. 23, indicate those parts of the body which may *not* have moved in the desired way about G, as G moves forward.

3. Explain the mechanical significance of the 'hang' style in long-jumping, whereby the jumper, leaving the board with some 'forward' angular momentum about the transverse axis, is able to make a satisfactory landing by immediately extending his body longitudinally as far as possible.
Explain why a bending at the knees should make the final piking action more effective in getting the legs up, and suggest an effective rotary arm-action during recovery after landing.

4. Show that, if R is to cancel the linear momentum of the vaulter in the situation of Fig. 26(b), the resultant of R and W must provide an impulse equal and opposite to $m \cdot v$.

5. Suggest reasons why, on landing from a gymnastic vault, a performer often recovers balance by a forward and upward swing of the arms as the body becomes erect.

6. Explain how it is that, in Fig. 26(a), the force R is changing the body's remote angular momentum about O into the local form about G, whereas in Fig. 27(b) it is doing the reverse.

7. A gymnast, swinging from a high bar, just reaches the

extremity of his swing with his body in full lay-out horizontally.

Show (a) that as he *starts* to swing back, the reaction of the bar on his hands is vertical; (b) that G's downward acceleration is therefore less than g; and (c) that, although bodyweight is causing a gain of remote angular momentum about the bar, this is being reduced somewhat because the reaction of the bar is transforming some of it to angular momentum of the local form about G as the body rotates.

8. Show that, if a backward somersault could be made before landing from the vault of Fig. 27, then, on landing, the line representing ground-reaction in Fig. 26(b) would need to pass beneath G.

What method of control can the vaulter use to determine his landing attitude? Should the fund of 'backward' angular momentum possessed by him make it easier to control the path of G as it is brought to rest?

9. Use the parallelogram construction to show that ground-reaction on the lower end of a pole during a vault is, like the big force exerted on it by the vaulter, directed almost along its length or along its chord.

10. It has been contended that force is associated with acceleration rather than being the cause of it. Does this contention remain valid when the fundamental physiological factors involved in muscle-contraction, and the forces exerted by muscles, are considered?

CHAPTER EIGHT

Mechanical Work and Energy

Energy has only been mentioned so far in connection with the changing condition of a system of bodies which, by mutual collision, eventually come to rest with respect to each other (p. 58). We have also noted that such internal action makes no difference to the linear momentum of the system (nor to its angular momentum about any axis), but that heat is produced as the motion of the bodies with respect to G diminishes. Heat in a material body is recognised as 'energy' in the form of random molecular motion; the collision between such bodies is therefore just a transformation of some of their 'gross' energy of motion—kinetic energy—to the equivalent measure of heat energy, the ultimate form of energy (Fig. 13).

Heat and mechanical work

The effect of force on material bodies has already been considered in a variety of situations, and the importance of the product of force and the time for which it acts has been made plain, particularly in connection with the laws of conservation of linear and angular momentum. However, equally important results follow from a consideration of the product of force and the distance through which its point of application moves in the direction in which it is acting: a physical quantity known as the 'mechanical work' done by the force. The importance of this is that when a force moves its point of application in this way, heat energy is eventually produced as a result; and the *same amount* is produced, no matter what other intervening processes there may be between the performance of the work and the production of the heat energy. A given measure of heat energy is always associated with an initial performance of a given amount of mechanical work, and is proportional to this product of force and the distance its point of application moves in the direction in which it acts.

If we define energy as 'the capacity to perform mechanical work', then it is easy to see the operation of another conservation law—the conservation of energy; for the intermediate processes between the performance of work and the final production of heat may themselves be regarded as operations whereby energy is changed from one form to another without loss. For example, consider the following sequence of events:

A diver climbs up to the ten-metre board and dives into the pool.

He climbs to the side of the pool again.

The disturbance of the water finally subsides.

In terms of energy transformation, the diver performs mechanical work in climbing to the board: an amount measured by the product of his body-weight and the vertical distance through which he has moved his centre of mass. This work has been done on his body, giving it 'potential' energy—by virtue of its high position—which, in turn, is transformed into 'kinetic' energy—energy of motion—as he loses height and gains speed in his dive. Entry into the water now reduces his kinetic energy—the energy of a uniformly-moving mass—in favour of the increasingly irregular whirls and eddies of the disturbed medium: a form of kinetic energy that eventually degenerates to the heat form. When the diver returns to the side of the pool again we know that whatever mechanical work he has done has all been transformed into its heat equivalent.

We will now realise that the agency responsible for the

performance of work (in this case the diver) will not be in the same condition at the end of the operation as he was at the beginning. Here, the work has been done by muscular action, and has been responsible for metabolic effects in the body—a loss of energy internally which will have to be made up later on. Furthermore, of the internal energy made available during the climb to the board, only a part is used in performing the necessary work; the rest is itself dissipated as heat.

In the above example we have chosen to start the sequence with the performance of mechanical work in the upward climb to the board. It now transpires that, in the activities involving human motion, when one form of energy is being transformed to another, mechanical work is the operative link between them. Also, the amount of work done, i.e. the product of force and distance involved, is itself a measure of the energy that is being transformed. The work done in the climb to the board, for example, is a measure of the chemical energy expended *usefully* in the body in providing it with an equal gain of potential energy. This, in turn, is transformed to kinetic energy when the force of gravity moves the body vertically downward to the water-level; and this is finally converted to increasingly random motion of the water by the force exerted by the body on the water over the distance the body travels in it before being brought to rest. Work is then done on each other by the various masses of moving water, to change this irregular movement into heat.

Resolution into components of force and distance   Very often, the force that is performing mechanical work is not moving its point of application in its own direction but in a direction inclined to it. The hammer-thrower of Fig. 4 (p. 20), for example, is at all times exerting force on the sphere along the length of its wire, but this is not the direction in which the sphere is moving: it is moving always along the tangent to its path. We must therefore resolve the force into two components: one at right-angles to the path of the sphere —directed towards the centre of its curved path, and therefore doing no work; the other, acting tangentially, urging the sphere in the direction in which it is itself acting, and thus performing work. This work is giving the sphere increasing kinetic energy, as shown by its increasing speed.

If, in a particular instance, the force has a constant direction, then it is better to separate the distance, rather than the force, into components. In the example of the diver, given above, the actual path of G will be part of a vertical parabola as he moves forwards and downwards from the board—a short horizontal distance covered as he drops vertically; but the force of gravity does no work horizontally because it has no component in this direction: all the work done is performed along the *vertical* component of distance, and is measured by the product of body-weight and the vertical change of position of G.

Energy a scalar quantity

Consideration of the various forms of energy, taking as a convenient example the kinetic energy of a body moving in a horizontal circular path, will show that 'capacity to perform mechanical work' is independent of any directional features of motion. This body has the same kinetic energy at all parts of its path, i.e. in whatever direction it may be moving; for, in being brought to rest, it is able to perform work in relation to its speed—not to its direction of motion. Evidently, mechanical work and energy are scalar quantities; and the fact that we can never measure the *total* energy possessed by any system of bodies means that we are only concerned with *changes* in this quantity.

The kinetic energy of a body due to the velocity of its centre of mass   We are already familiar with the equality of linear impulse and the resulting linear momentum it gives to a body originally at rest. This has been expressed in f.p.s. and SI units, respectively, by:

$$32 . F . t = m . v \quad \text{and} \quad 9{\cdot}8 . F . t = m . v$$

as on p. 54, where F is given in lb wt and kg wt, and v, the velocity of the body's centre of mass, necessarily has the same direction as F. The distance, s, through which the force moves the centre of mass in the time, t, can be introduced by using the fundamental relation:

$$\text{distance} = \text{mean velocity} \times \text{time}$$

If F is assumed to remain constant, so that the body gains its velocity uniformly with time, then the mean velocity from rest is half the final one, v, and we have:

$$s = \tfrac{1}{2} \cdot v \cdot t$$

which, combined with the equations above, gives:

$$32 \cdot F \cdot s = \tfrac{1}{2} \cdot m \cdot v^2 \quad \text{and} \quad 9 \cdot 8 \cdot F \cdot s = \tfrac{1}{2} \cdot m \cdot v^2$$

for the f.p.s. system and SI, respectively.

This result has been obtained on the assumption that the body has moved *freely* from rest, and has therefore gained no other form of energy, nor lost any by doing work itself. If, before having work done on it, the body were moving with velocity u, and afterwards with velocity v, then the work performed on it would be responsible for a gain in its kinetic energy of $\tfrac{1}{2} \cdot m \cdot v^2 - \tfrac{1}{2} \cdot m \cdot u^2$ (the simple arithmetic difference between the quantities), no matter how the direction of motion of the body changed during the operation. This is so because, as noted above, these are scalar quantities, for which directional attributes do not exist.

Units of work and energy   A product of force and distance has already been found in the measurement of the 'torque' due to a force about an axis (p. 24); but, in that case, the operative force acts at right-angles to the distance between its line of action and the axis, and also at right-angles to the direction of the axis. This is the 'vector' product of force and distance, and is usually expressed in its practical units of lb wt-ft or kg wt-m. In the f.p.s. system, the corresponding unit of work is differentiated from that of torque by being given the form ft-lb (an abbreviation of ft-lb wt). This is convertible to the absolute unit, as above, by the use of the factor 32, so that $32 \cdot F \cdot s$ is the measure of work or energy in ft-poundals.

Similarly, $9 \cdot 8 \cdot F \cdot s$ (with F expressed in kg wt) is its measure in what could be called 'metre-newtons'; this unit is in fact known as the 'joule'.

Energy considerations in the use of resilient apparatus

Energy degeneration from the kinetic form to heat is, as suggested by the examples quoted, a universal phenomenon with an inevitable finality. It is one that limits performance in every form of human physical activity, for it always arises when the body moves through a fluid medium, when it slides over a surface, or when it comes into contact with any other material object. It comes into operation, for example, whenever a body undergoes a change of shape—a deformation such as that experienced by two bodies in collision. As we have seen, the ability of such bodies to preserve a high proportion of their original kinetic energy, and to move away from each other nearly as fast as they approached, depends on their ability to exert on each other, at all stages of their recovery, almost as big a mutual force as was experienced by them both in the process of deformation, when kinetic energy was being partly transformed into energy of strain.

Probably the most important practical examples of collision phenomena with which we are concerned are those involving the rapid and repetitive contact made between the human body and the ground in walking or running: contact which is essentially that of a collision in which we hardly expect any kinetic energy to be preserved. This would certainly be the case if the reduction were not made good, at every stride, by vigorous muscular effort by the body itself: an attempt to develop a kind of 'artificial' elasticity where little exists in either the foot or the ground. It is to assist muscular action under these circumstances that resilient devices such as 'springy' boards and trampettes are used for take-off over high gymnastic apparatus; and why trampolines can help to build up big vertical speeds in the gymnasts using them. Even the feebly-resilient properties of modern track materials have a beneficial effect on running performances, and we have already noted the near-perfect elasticity of the fibre-glass pole in providing the vaulter with almost all the strain energy stored in it as it recovers with nearly the maximum possible speed.

Consider the behaviour of the feebly-resilient structure of a gymnasium floor under conditions of collision with the downward-moving foot of a gymnast. Its immediate response depends very much on the kind of contact that is made; for, if the man attempts to keep rigid, his whole body will be brought to rest quickly and the region of the floor around the contact-point will suddenly be forced to accelerate rapidly downwards—so rapidly that the disturbance might be of the nature of a 'shock': a big force acting upwards on the foot and downwards on the small area of the floor around it, but acting for a very short time. The subsequent effect will be the rapid spreading of a system of complex up-and-down vibrations through the material of the floor—a series of shock waves which deliver numerous upward impulses of diminishing intensity to the body of the gymnast, and reduce his momentum to zero before there has been any *general* depression of the floor surface itself. Needless to say, the amplitude of the disturbance is very small, but, together with the similar one dissipated within the gymnast's body, it accounts for nearly all the kinetic energy expended by him.

In this, we recognise the same kind of change that was noted above in the case of the diver (p. 122): a change from the 'gross' kinetic energy of the uniformly-moving body to an increasingly random motion, this time in the form of undulations spreading through both body and floor, and soon to be degenerated to the form of heat. We realise, also, that the energy in its 'random' form of motion cannot be returned to the body as kinetic energy of upward movement—the form in which it might be required—and muscular action by the body itself will be needed to make good the loss if anything like a 'bounce' is needed.

The possibility of introducing muscular effort in order to develop a 'bouncing' contact suggests that a skilful performer can reduce or eliminate landing-shock on any resilient surface by, for example, landing on the ball of the foot and flexing the leg at ankle, knee, and hip so that at the instant of contact there is little or no downward movement of the foot—a delicate foot-placement which can then, by further control of the limbs, be made to develop a comparatively slow downward acceleration of the whole landing surface. This produces no vibrational impulses and no attendant energy-losses: just a steadily-increasing upward force on the foot, its effect transmitted to all parts of the skeletal structure, and controlled by its associated muscular system.

In practice, of course, the very vigour of most gymnastic efforts makes impossible the attainment of zero relative speed between the foot and the resilient apparatus; the design of the latter should therefore be such as to minimise vibrational losses. These hardly exist in a device such as a trampoline, the general deformation of which is much more evident than the local vibration of its bed, and ensures only a slow retardation of the performer's body under a force never big enough to impose excessive stress on any part of it, or to cause big local variations in its speed. There is always plenty of time for the steadily-increasing upward force to develop the impulse which brings G to rest at its low-point; there is, likewise, time for the same upward force to start G moving upward again, giving the body more and more momentum until it loses contact.

It should now be evident that there are two principal ways in which collision phenomena in physical activity lead to loss of performance, and that even if the direct degeneration of energy via local disturbance is minimised, the physical properties of the ground, or of the apparatus involved, are such that the force it can exert during recovery from deformation is at every stage less than the corresponding force required initially to deform it. The energy of strain stored in the deformed apparatus is never quite as much as the kinetic energy to which it should, ideally, be equivalent, and can never completely be returned to the kinetic form again. This is something entirely out of the control of the performer: it depends entirely on the elastic properties of the apparatus, which should allow only a slow energy-degeneration to occur. If a trampolinist wishes a series of bounces to be of constant height, he must arrange for all the kinetic energy he loses as his centre of mass is brought to its low-point, to be given back during further contact with the trampoline. Only an intelligently-timed posture-change by the performer can do this: one which, in its simplest form, allows the legs to become flexed as G approaches its low-point, so that they can be vigorously straightened when G arrives there. The increased depression of the trampoline bed now gives rise to an increased upward force on the feet, an increased upward acceleration of G, and a final kinetic energy of the body either greater than, or at least equal to, its initial value.

Looking at this in more detail from the point of view of energy, we see that in the first phase the performer's kinetic energy of downward motion is mostly transformed into energy of strain in the apparatus as G's speed is reduced to zero; but that losses due to imperfect elasticity make it impossible for all this strain-energy to be returned to the body in the kinetic form as the depressed surface recovers. The energy deficiency is made up by drawing on some of the body's own internal energy, made available in this instance by rapid leg-extension, and changed immediately from this kinetic form into the requisite extra energy of strain in the further depression of the trampoline bed. The operation is finally completed by the transformation of most of this into the kinetic energy of the upwardly-driven body.

Energy Transformations in Swinging Activities

Features of pendulum motion

The principles which govern the motion of gymnasts swinging from horizontal bars or from rings are those applicable to pendulums generally, and can be illustrated conveniently by the use of the following simple pieces of apparatus (Fig. 30):

(a) A small compact body such as a small lead weight or a heavy bead, A, capable of being regarded as a 'particle', hangs from the end of a piece of thread some 2 ft long. The free end of the thread passes through a small hole drilled in a ruler held rigidly over the edge of a table, as shown, to form a simple pendulum. The length of the pendulum can be varied within limits set by the two fixed stops, B and C, fixed to the thread 2–3 in apart.

(b) A large empty wire-bobbin or drum, A, is mounted on a horizontal axle which is, in turn, supported in a rigid framework one or two feet long, and is itself capable of swinging about a similar axle fixed at B. One end of a length of thread is fixed to the drum so that the thread can be wound round it a few times and then taken up through holes in the framework to pass over the axle at B. A pull on the free end will cause the drum to rotate in one sense: it will reverse its rotation when the pull is relaxed because there is a long light elastic band wound round it and fixed at C, to rotate it back to its former position. This is also a pendulum, but its mass is distributed throughout the configuration of the drum and the framework. It is a 'compound' pendulum, designed to simulate the action of a gymnast hanging from a bar and attempting to start a swinging motion from rest, using the 'cyclic' rotary action of the legs.

(c) This is as (b), above, but the rotary motion of the drum is replaced by the lever-arm action of a short wooden lath hinged to the vertical framework near one end. This is actuated, also, by the thread, to simulate the piking action of the gymnast.

It is evident that at the extreme points of the oscillation of the simple pendulum (Fig. 31(a)), where the velocity is momentarily zero, the system has potential energy only—energy due to the high position of its centre of mass. As G falls to lower levels, more and more energy of this form is changed to kinetic energy; and this will have its maximum value at the low-point, L, where no further transformation of

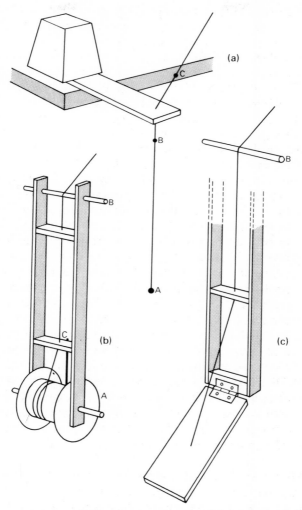

Fig. 30. Models for investigating pendulum motion.

potential energy is possible. In the absence of losses, this kinetic energy will be expressed as:

$$32 \, . \, W \, . \, s = \tfrac{1}{2} \, . \, m \, . \, v^2 \quad \text{or} \quad 9{\cdot}8 \, . \, W \, . \, s = \tfrac{1}{2} \, . \, m \, . \, v^2$$

where W is expressed in lb wt or kg wt, respectively, and s is the vertical distance through which the *vertically downward* force, W, has acted.

Again, if no energy losses occur, the kinetic energy of the 'bob' at L will be reconverted to potential energy at the same high level; but, in practice, this level is never attained again, and the oscillations are damped out finally with the bob at rest at L. Experiments with the model of Fig. 30(a) will show that if the centrifugal pull and the weight of the moving bob on the support cause the latter to be depressed noticeably, then loss of energy takes place relatively fast. This is because the extra potential energy, made available by the slight fall in L, is partly stored as energy of strain in the flexible support, and this is not all returned to the pendulum when, at the extremities of the swing, the support recovers.

As with the trampoline and other apparatus discussed earlier, energy losses have constantly to be made good by the performance of more mechanical work. In the pendulum model this work is done via the thread; in the case of a gymnast swinging from a bar, it arises from muscular effort associated with body-metabolism. If the amplitude of the oscillation is to be maintained, this work must be done on the pendulum when maximum force is needed to perform it. If the simple pendulum model is set swinging from its upper stop, C, for example, with an amplitude of a few inches, this can be increased by shortening its length by as much as the lower stop, B, will allow, every time the bob passes L (Fig. 31(b)), and only allowing it to extend again at or near its high-points. This technique ensures that the mechanical work done *on* the pendulum when its bob is raised by the shortening thread at L is more than the work done *by* the bob when it extends the thread by the same distance near the extremity of the swing; for, at L, the force needed is the *sum* of the weight of the bob and the centripetal force that is keeping it moving in a curved

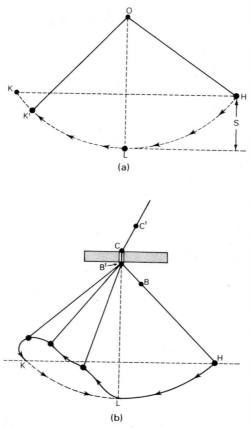

Fig. 31. (a) If, in a pendulum-swing, energy exchange between the potential and kinetic forms were complete, then the centre of mass, starting at H, would reach the same level at K. In practice, it only reaches K′.

(b) In Fig. 30(a), energy-loss is made good by work done *on* the pendulum as it moves near its low-point, L, where its length is shortened by the distance between the stops B and C.

It is only allowed to *lose* energy near the extremity of its swing, where it exerts only a small force on the thread in regaining its original length. The operation can be repeated indefinitely.

path at its maximum speed; whereas, at the extreme points of its swing, the bob has zero speed around O and is able to exert only a fraction of its weight along the length of the inclined thread. As a result, it is easy to build up a big amplitude, just as the gymnast does when he pikes slightly, or raises his centre of mass by other means, as he passes his low-point, only allowing full body-extension at some other part of his swing.

The swinging of an extended body   The motion of a simple pendulum—a localised mass—has only limited application in the study of swinging activities of the human gymnast—a *distributed* mass. The essential difference, illustrated by the difference between the model shown in Fig. 30(a) and those in (b) and (c), is that the extended body can have rotary motion, 'local' angular momentum, and kinetic energy of rotation (p. 142) about its centre of mass, and thus can initiate a swinging movement *from rest*: a facility denied to a simple pendulum, which can only increase an amplitude of swing already given to it. A change of posture, such as the piking movement referred to above, not only brings G nearer to the support but constitutes an increase of local angular momentum about G which can be developed more effectively near the extremities of the swing than near the low-point, as experience soon shows.

Interpretation of specific forms of body-movement may, evidently, be presented in terms of energy transformation or by considerations involving impulse and momentum, the choice depending on the nature of the movement and the aspect of it that has to be studied. Where a rotary impulse about a fixed origin is changing in a complicated way as the action goes on, as in pendulum motion, the use of the scalar quantities may be recommended; but where the motion is dependent on the rotary effects of posture-change, we find a more obvious explanation in terms of rotary impulse and change of angular momentum. For the pendulum models of Fig. 30(b) and (c), and for the corresponding body-movements they represent, both treatments have their place—

energy transformation to explain the basic form of the motion, and rotary effects to show how it is initiated, built up, and controlled.

The behaviour of (b) may be likened to that of the turntable in Experiment 3 (p. 85) when a rotary impulse, delivered to its drum by a pull in the thread, gives the drum *local* angular momentum about its axle; for this simultaneously confers on the whole pendulum an equal and opposite measure of *remote* angular momentum about the axle through B. This impulse has no turning-moment of its own about B because the thread is either taken round the axle, or, ideally, passes up through a hole bored in it. The immediate effect on the static pendulum is to start it rotating about B with zero angular momentum, i.e. to displace its centre of mass horizontally while the drum rotates. This effect is basically the same as the displacement of G in the stepping-forward from rest (p. 95); but whereas in that movement the body continues to lose balance, under gravity, in the same direction as G's displacement, here we have the force of gravity, acting through G, developing an increasing rotary impulse about B in the sense *opposite* to that required to keep G moving in its displaced direction. As in all pendulum movement, G is soon brought to rest, and has its motion reversed, as this rotary effect of the weight removes and reverses the pendulum's remote angular momentum about B; and this happens even if the drum is kept rotating. In fact, even to keep G at rest in a displaced position the drum's rotation would need to be constantly accelerated, for it would have to supply local angular momentum to the system at a rate equal to that at which the remote form is being steadily removed from it by the torque due to its own weight.

Rotary acceleration of the drum, in one sense or the other, is evidently the controlling factor in the build-up of a bigger amplitude from the small initial displacement. All that is necessary is for its timing to be correct: the acceleration must cease as soon as G comes momentarily to rest, at which point the spinning drum must be retarded to rest, the retardation helping to reverse the motion of G and send it back fast through its normal hanging position. Acceleration of the drum in the opposite sense can start at any time after this, but it must continue until G has come to its other extreme displacement: then, and only then, can retardation be started. In practice, the limited range of rotation of the drum makes it necessary to control its changing rotary speed when it is in the neighbourhood of the extremities of the swing; for it has usually been brought to rest soon after leaving the extreme positions, especially when a big amplitude has been generated.

The gymnast can test the technique by hanging from the high bar and using the cycling action of the legs—an accelerated rotary movement, first in one sense and then in the other—to initiate a swing and to build it up: he will soon appreciate the importance of timing. He can also experiment with the more natural piking action (embodied in the working of the model in Fig. 30(c)) where the local rotary features of the method described above are accompanied by a raising and lowering of G, as with the simple pendulum in (a). It should be noticed, too, how much closer is the action of a child on a 'swing' to the rotary motion in (b) than to that in either (a) or (c).

Problems

1. Develop a high bounce from the bed of a trampoline, noting when the vigorous leg-extension has to be made in order to do so. Discover, then, what has to be done to 'kill' the bounce, and remain in contact with the trampoline. Explain the technique.

Use a rapid flexing and extending of the legs continuously as contact is made, to simulate the production of vibration in one of the bodies involved in a collision. Explain the result obtained.

2. Perform some sprinting experiments at maximum steady

speed: (a) on a cinder or grass track, using spiked shoes; (b) on a horizontal stretch of loose sand. Measure stride-rate and stride-length in the two cases, and find out which is more affected by the adverse conditions of (b).

It may now be possible to decide whether energy-loss is due mainly to the depression of the sand (causing reduced foot-speed and stride-*rate*), or to a forcing back of the sand (causing a reduction in stride-*length*).

3. Consider, in terms of possible reduction in the energy lost when foot-contact occurs in sprinting, whether or not the use of 'springy' brush spikes is likely to give a runner an advantage, particularly in the accelerating phase of a race.

4. Try to explain the build-up of amplitude of the simple pendulum of Fig. 31(b) in terms of changes in angular momentum about its support.

5. Suspend a bicycle ergometer, or similar device, by ropes hanging from the ceiling. By driving its wheel, first in one sense and then in the other, discover the appropriate times when the wheel must be accelerated and retarded to build up a big amplitude of swing about the support. Explain the result in connection with Fig. 30(b).

6. What should be the mechanical properties of a 'properly-sprung' dance floor?

CHAPTER NINE

The Utilisation of Effort

Work done by a varying force over a fixed distance

Every propulsive effort, whether it is made in discharging a missile or in producing acceleration of the body itself, employs vigorous trunk and limb movement carried out by the appropriate muscle-groups: and these movements are clearly limited by factors of distance rather than by those of time. Where only the magnitude of the result is under discussion, therefore, it is more profitable to examine such effort in terms of mechanical work rather than from the standpoint of impulse and momentum.

If we are investigating principles only, and not the complex details of a particular action, then we can base our thinking on

Fig. 32. Energy-losses in limb movement: (a) and (b), graphical representation of the work done by a constant force, F, and by a force diminishing with extension, s.

In (c), the horizontal force, F, exerted on a missile, A, is less than the force R exerted by the fixed shoulder because the arm-components have to be accelerated in the directions shown at G_1 and G_2, and given rotation about these points. The weight of the arm and missile must also be supported. (d) shows that the elbow, E, of a straightening arm has to be *accelerated* through positions 1 to 5, even when the extremity is not accelerating.

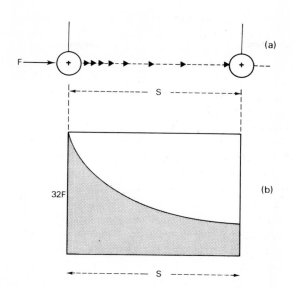

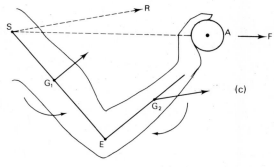

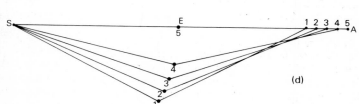

the behaviour of a suitable conceptual 'model' with which mechanical work can be considered to be done, first under ideal conditions and then under conditions which can be made gradually more realistic. As in Fig. 32(a), we imagine any device capable of exerting a constant force, F, through a horizontal distance, s, on a stationary body of mass m. The body can be supported, supposedly, by a very long thin wire, so that its weight does not have to be considered, and, in the absence of other extraneous effects, the mechanical work done on it will be equal to its kinetic energy. Over this fixed distance, s, the body's final velocity, v, would be higher if F were made bigger, in spite of the fact that F would then be acting for a shorter time.

It is now easy, by using the graphical method of Fig. 32(b), to apply this result to the case where F is *not* constant; for if the rectangular axes represent force and distance in, for example, f.p.s. units, with F expressed in lb wt, then the product $32 . F . s$, or $\frac{1}{2} . m . v^2$ ft-poundals, is seen as the area of the unshaded rectangle when F is constant and that of the shaded area if F diminishes as s increases. From this it is clear that the greater the enclosed area, the higher will be the final speed given to the projected body; and we can argue that in any action in which we do not know the precise way in which the force is going to vary, maximum speed will still be achieved when the greatest possible force is exerted at all stages of the action in the direction in which the motion is taking place.

Energy losses in limb-extension

The example of a driving-force diminishing as the distance of its action increases can be realised in practice by a model of a compressed spring which drives the body horizontally before it as it extends from rest. Such a model would also show that when the spring is given a less massive body to accelerate horizontally, its speed of extension is necessarily higher at all stages of its movement; but, on account of its own mass, the *force* it is capable of exerting on this body is less than it could exert before. This accords closely with the behaviour of an extending limb performing work by muscular contraction (Fig. 32(c)); for when the limb drives a body forward in this way, the force actually experienced by the body becomes less as both the distance and the speed of extension become greater; furthermore, the work done on it is always less than that made available by muscular action. The system behaves just as a machine behaves—wasting a proportion of the mechanical work done on it by the contracting muscles, the lost proportion rising to 100 per cent when the limb can extend no faster, and can therefore give no acceleration to the body.

In common with the spring-operated model, the masses of the moving limb-components introduce energy losses which reduce the amount of available work the limb is capable of doing externally. These losses are simply due to the fact that when the body is being accelerated there must be acceleration of the individual components of the limb as well, and these accelerations vary in magnitude and direction in a way which depends on how the limb-extension is being carried out. Fig. 32(c), for example, shows that not only has the weight of the extending arm to be supported, but the reaction, R, of the stationary shoulder has to provide for the upward acceleration of its two components also, as the mass, A, is driven from rest horizontally. The vertical part of this reaction is not used at all as far as A's motion is concerned, and some of its horizontal component has to be used to accelerate the arm this way, only leaving a residual force, F, to give horizontal acceleration to the mass, A. It is important to realise that even this force can be reduced to zero before full arm-extension occurs unless wrist and finger movements are introduced; this can be shown with the jointed-rod model of Fig. 32(d), which represents the humerus, SE, and the forearm and missile, EA, as in (c), and portrays their changing positions as the arm becomes straight. If S, the shoulder, is kept fixed, and the elbow, E, has the successive positions 1 to 5; then the extremity, A, will be found to be moved through its own successive positions; but here these are equally spaced, whereas those of the elbow show rapid increase in displacement, especially when the arm is nearly straight. The changing geometry of the arm makes it necessary for its main components to rotate more and more rapidly round their joints *even to keep the extremity moving at a constant rate*—let alone to accelerate it. The force driving the missile forward has therefore been reduced to zero well before the arm becomes straight, and this situation can only be remedied by using forward wrist-and-finger acceleration to maintain the force; or, as will normally be the case in practice, the shoulder must itself be accelerating forward, driving everything before it.

Energy expenditure in a 'straight' arm blow   Although the arm, extending under the conditions discussed above, can produce no acceleration of its fast-moving extremity when it is nearly straight, its two components have each had plenty of mechanical work done on them, and possess considerable kinetic energy. If, therefore, the object of the action is not to develop high speed in the missile but to deliver a blow with the clenched fist, as in boxing, then the kinetic energy of the straightening arm can be purposefully expended. The arrows (Fig. 32(c)) show the existence of this energy in the individual velocities of the two centres of mass, G_1 and G_2, together with the energy of rotation of the components as they change direction around G_1 and G_2 in the senses indicated. The sum of all these quantities is available for performing mechanical work against any opposing resistance-force which can bring the arm-components to rest as, or before, the arm becomes straight.

If we assume S to be fixed, then the extremity of the arm, A, can only be brought to rest under these conditions when the nearly straight arm is made to expend its kinetic energy by doing work, and this in the exertion of a very big force over the comparatively short distance available to it between the numbered positions of A (Fig. 32(d)). It must be noted, also, that the available kinetic energy becomes greater as the faster-moving elbow approaches its final position, E.

It must also be remembered that the force exerted by the

fist in the delivery of a blow is equal and opposite to that sustained by the shoulder, which will tend to be forced back, so reducing the work done by the fist. Momentum considerations show that the loss will be small provided that the upper part of the body is much more massive than the 'localised' part of the opponent's body, e.g. the chin, to which the impulse is delivered.

The use of several extending components

The above discussion shows that an attempt to accelerate the extremity of a limb is severely limited, particularly in the final stage of its extension, if the other end of the limb is kept static. The distances through which the extremity moves in equal times under these conditions do not increase; they may be equal, as suggested in Fig. 32(d), or even diminish as the limb becomes straight.

We now study the principles involved in the use of two or more extending components in their effort to perform more mechanical work on a missile. The increased propulsive effect of a series of such components lies in their ability not only to widen the range of the action—i.e. to increase the distance s (Fig. 32(a))—but even to apply to a projected body a force, F, greater than a single component can exert. This feature is bound up with the serious reduction in the useful driving-force of an extending limb as the speed of its extension increases, and to examine this we need to consider the working of a more elaborate spring-operated model, as shown in Fig. 33(a).

Here, in addition to the spring-type extensor, B, in contact with the sphere, A, and extending through a distance s_1, there is another one, C, capable of driving the whole system horizontally away from a fixed support, D, through a further distance s_2. If we regard B as something corresponding to an extending arm in a throwing action, then C's extension could represent the movement of the more massive shoulder, trunk, and hips in the same direction. If we first suppose that C stays at rest while B extends, accelerating A before it, then the

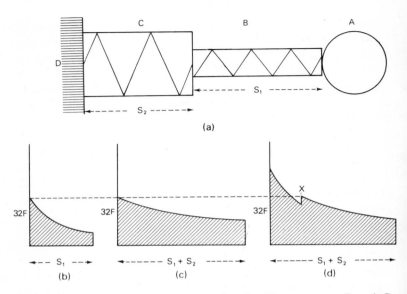

Fig. 33. The model, (a), consisting of spring-like extensors, B and C, simulates a throwing action as it drives the mass A towards the right, away from the fixed support D.

Force-extension diagrams for the work done on A would be expected to resemble (b) when B extends by itself, (c) when B and C start and finish their extensions together, and (d) when C's forward acceleration is initially so great that B cannot start its extension until X has been reached.

The discontinuity at X could arise because, when explosive action starts, A's acceleration becomes greater than that of B. The shaded areas are proportional to A's kinetic energy in the three cases. F and s are expressed, as before, in lb-wt and ft, respectively.

force-extension diagram will be one like Fig. 33(b), showing the diminution of the available force, F, as the speed, and therefore the extension, increases. As seen earlier, the area enclosed by the graph and the two axes represents the kinetic energy given to A.

Consider now the improvement if, while this extension is going on, the extensor, C, is also in operation, accelerating everything before it. This not only makes the total extension greater ($s_1 + s_2$), but it makes B's own extension-rate slower than before, and because of this it is able to exert on A, at all stages of its extension, a force greater than it could exert without C's accelerating help (Fig. 33(c)). It must be realised that C cannot exert force *directly* on A: the greatest force it can transmit to A is that which B can exert on A; and this depends on the 'strength' of B and on its rate of extension. Again, the force-extension diagram refers to A only, the kinetic energy acquired by B and C being a function of their masses and their distribution of mass as they extend. (Compare the practical case of Fig. 32(c).)

It has been assumed that both extending components, B and C, start and finish their extensions simultaneously, but under conditions of maximum performance this may not be desirable. Suppose, for example, that the more powerful component, C, is able to exert on B a force much greater than has been suggested: great enough to prevent B's extension from starting, so that it cannot drive A away from it. Under these conditions C will be accelerating forwards, and both A and B will be pushed forward with the same acceleration at the end of it: there will be no separation of the components—no 'explosive' action. This, however, is a situation which, in a throwing action, cannot last indefinitely because the powerful thrust by C becomes weaker as the speed increases, until (at X, in Fig. 33(d)) it becomes equal to what B can exert in extending. B then drives A to its full extension, and, if both extensions finish together, the work done on A over the full range of movement is even greater than before.

The above example shows how the accelerated extension of one component can delay and retard that of another, and so cause it to do more work on a missile. It is obviously desirable for maximum force to be exerted on the missile over the full range of available distance, and this emphasises the need for all major components to finish their extensions together; or, if this is not possible, for the more remote, massive and slow-moving ones to finish first. It is clear, for example, that if, in our model, B reaches its limit before C has done so, then it will suddenly lose speed, and contact with A will not be maintained; but if C stops moving, then although B will be partially checked, its extension-rate immediately increases and it can still exert force on A. Even this adverse effect can be minimised if C, having reached the limit of its extension, is free to continue moving bodily forward; for B is now able to proceed with its own extension, exerting force in the forward direction on the missile and backwards on C. This transfers some of the latter's momentum to the missile: a process that has 'recovery' value in competitive throwing. The above principles should now be applied to the conditions obtained in practical cases, where, in addition to general forward body-movement, there is final wrist and finger action available to the thrower.

Explosive action in jumping

A distinction has to be made between the 'explosive' effort required for the projection of a missile and that designed to give a high speed to the body itself. Although it can be argued that maximum performance in throwing is itself concomitant with a high body-speed, the link is not fundamental, but one resulting from the fact that maximum performance in most physical activities requires the co-operation of all parts of the body.

In thinking about the development of a big vertical force from the ground, we have to examine the way in which the driving-components of the body are disposed for this purpose: and here we see that the more massive, slower-moving parts such as the hips, trunk, and shoulders, which are centrally placed, have the capacity to exert on the more slender, fast-accelerating limbs a very big force over comparatively short distances. This is generally achieved in high-jumping by some form of rotation involving twisting or flexing of the trunk; but the important thing to note is that the presence of these central components may well prevent immediate extension of the more heavily-loaded lower limbs. This is because these not only carry the static weight of everything above them, but because they also provide the means of evoking from the ground the very big force which accelerates all these parts upwards—a force sometimes reaching three or four times body-weight, and usually transmitted by one leg. One result of this is that explosive action—the simultaneous extension of individual parts of the body—is initially confined to the upper body only. If, for example, a standing vertical jump is undertaken from the usual flexed-leg posture, then it is possible that the straightening of the trunk, with head, shoulders, and arms all accelerating upwards, may react downwards on the legs with a force big enough to cause them to flex still more. Only when the inevitable reduction in the upward acceleration has occurred will it be possible for extension at the knee, and then at the ankle, to make their final late contributions to the body's upward momentum.

Eccentric contraction   The factor that has most to do with the capacity of a limb to resist an enforced flexure, as in the case of the jump, and then to perform mechanical work in subsequent extension, is the well-known physiological one whereby a muscle undergoing 'eccentric contraction' (a contraction in which its attachments are being driven further apart) is able to resist the stretching force by developing a restoring tension much greater than it can exert either isometrically or in the process of doing mechanical work.

The important thing, from the mechanical point of view, is that when a muscle has been put into this condition and then immediately allowed to perform work *concentrically*, the force it is capable of exerting externally is a good deal greater than it could otherwise develop.

It is clear, therefore, that an initial rapid explosive movement by the central body-components and the arms has the essential function, in a jumping action, of conditioning the muscles of the legs for the big force they are immediately due to exert. If the body is allowed to drop a short distance before ground-contact is made, then an even more desirable result is obtained. If this is supplemented by using a short run-up to a chosen point, and a take-off after using the jumping leg to check forward speed, as in the high and long jumps, a still more satisfactory performance is achieved.

Control of Ground-reaction

So far, it has been taken for granted that the posture-changes required for a particular action can be carried out so as consciously to derive from the ground the kind of reaction needed for it. In the case of a take-off from apparatus in vaulting, for example, the characteristics of ground-reaction are determined by the controlled co-ordination of movement of all parts of the body taking part in it: only thus will the resultant impulse have the required magnitude and direction.

In other examples we find that the ground-reaction on the 'driving' leg is determined not so much by the muscles of this leg as by the controlled acceleration of the other one, as in the step forward from rest (p. 95). However, a fairly clear distinction can be seen between this type of physical effort, wherein the driving leg is never under very severe stress, and one in which it is suddenly called upon to accept a compressive thrust of such magnitude as to force it to behave after the manner of a stiffly-braced, springy strut: one which flexes and extends again entirely in accordance with the externally-applied force, and is capable of imposing its own characteristic extension-force only when conditions allow it to do so. Clearly, in this case the force-pattern from the ground is almost entirely determined by what other parts of the body are doing.

This situation arises most often when the body is passing quickly over the grounded foot, giving only a short time for ground-reaction to develop the required impulse. It is characterised by comparatively little leg-flexion, unlike the big angular displacements of the leg-components seen in vaulting, where the use of resilient apparatus gives much more contact-time to the gymnast and permits a *controlled* leg-extension to drive him from it. There is never time, on the other hand, for the fast-moving long-jumper to gain from the board the upward impulse needed for the optimum flight-path of his centre of mass; nor does the stiffly-braced grounded leg have much controlling influence over it.

Direction of the impulse from the ground

Some of the basic principles concerning the control of ground-reaction have been discussed in connection with recovery of balance, and allied movements, in which the desired effects are produced by rotary acceleration of parts of the body about G, and not by any conscious action involving the grounded leg. Now we attempt to see how the controlled extension of this leg does determine the magnitude and direction of the force it derives from the ground in those cases where it is primarily responsible for doing so.

As in other aspects of human movement, the principles are clear-cut but their application introduces extraneous effects. If, from the usual flexed-leg posture used at the start of a standing jump, one were to drive vigorously against the ground by a series of leg-extensions—first at the hip only, then at the knee only, and finally at the ankle only—the widely-differing results would show convincingly that the desired direction of the force from the ground is achieved only when these efforts are made simultaneously and have the correct relative magnitudes. Accurate analysis, however, is made impossible by the existence of two-joint muscles, which prevent the complete isolation of these effects from each other. Furthermore, to investigate even a static posture, account should be taken of the individual weights of the leg-components, and, when accelerated extension is going on, their radii of gyration and other rotary features must be known.

The matter can be examined in its simplest form with the help of another jointed model (Fig. 34), capable of representing flexion and extension about the hip, H, the knee, K, and the ankle, A. Springs or other devices can be used to activate the model, if desired, and it is useful to be able to 'stiffen' the ankle-joint when this is needed to prevent rotation of the foot around it. A peg through H can be used to pin the model to a baseboard. Simple tests can now be carried out to show that a fixed length of string connecting A and H will entirely prevent extension about K without interfering in any way with the model's freedom of rotation about H. This enables us to see that if it is possible for muscular action to take place at the knee only—to cause a 'drive' against the ground while no effort is made at either the hip or ankle—then the bearing-surface of the foot against the ground will be at the heel, and ground-reaction, R, equal and opposite to the 'drive', will be directed from this region up through the hip (Fig. 34(a)).

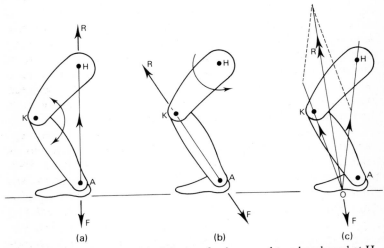

(a)   (b)   (c)

Fig. 34. A pin-jointed model of the leg, freely pegged to a baseboard at H, can be used to show how action at H, K, and A determines the direction of ground-reaction. Extension at K only, as in (a), can be prevented by the cord linking H to A. Resistance to rotation of the thigh only, as in (b), can be achieved by a pull in the cord from A through K, the knee.

The parallelogram construction, as in (c), gives the resultant ground-reaction, R, when the two actions operate together. The origin, O, is determined by the rotary effort being made at the ankle, A. The 'drive', F, against the ground is shown in each case.

It is obvious that a force from the ground for which the hip, rather than the knee, is responsible, must be one directed straight *through* the knee; this can be confirmed with the model by attaching the end of the string to A, as before, and pulling along the line AK. Extension at H can be enforced against this pull, but KA simply behaves as a pin-jointed strut, and maintains its direction (Fig. 34(b)). Individual action at the hip only, then, would bring forth ground-reaction along AK; and the resultant of the two extensions, acting together about hip and knee, would be found by using the parallelogram construction, as in Fig. 34(c). This diagram also shows how ankle-action and extension of the foot about it can enter into the argument. The discussion involving Fig. 11 (p. 48) shows how a redistribution of ground-reaction over the bearing-surface of the foot can control the point at which its resultant acts, and so make possible the maintenance of static balance. This redistribution is effected by slight extension or flexion of the foot about the ankle. Ankle-action, then, in conjunction with action at hip and knee, gives rise to a parallelogram of forces having its origin somewhere between the heel and the toes, the precise location moving away from the heel as the turning-effect at the ankle increases.

The modifying effect that the weights of the leg-components have on the direction of the forces involved in its extension can be seen if the model is mounted with its baseboard vertical. The pull in the string will now be found to pass slightly *above* the knee, K, to keep the lower leg in equilibrium, for a small part of its vertical component is needed to support the weight of this part; and if, in extending, the leg-components accelerate in any direction, then part of ground-reaction must supply the necessary forces. As will be seen from Fig. 34(c), the resultant normally passes up from the foot and then *between* H and K, being nearer to the joint developing the greater turning-moment; but under conditions of great acceleration, e.g. in the sprint start, it can be directed above the hip as the forwardly-directed legs straighten.

Mechanical Features of the Running Stride

Forces acting on the runner

To discuss this topic it is necessary to distinguish between 'sprinting', wherein energy-economy can be disregarded in favour of the attainment of maximum speed, and 'distance-running', in which such economy is of major importance. The line of demarcation between these is a matter of opinion, and is moving steadily towards the longer races as time goes on. Slow-motion films show the distinction clearly in the stride-pattern associated with each form of running, some features of which are illustrated in Fig. 35. Films also establish the fact that as speed and stride-length increase during acceleration from rest, the proportion of the time spent in contact with the ground decreases, so that at the final steady speed of the distance-runner the stride-rate may be of the order of three per second, with only half the time spent on the ground. For top-class sprinters the figures are around $4\frac{1}{2}$ strides per second, with only about 40 per cent of the time spent in ground-contact.

We have already encountered the difficulty of deriving from the ground a reaction having a pronounced forwardly-directed component; this is seen particularly in films of the early stages of a sprint race, where (as in all activities) forward acceleration depends on the maintenance of continuous contact with the ground, but where ground-reaction is constantly driving the runner from it. The vertical component of this force on the body of the runner becomes even more pronounced as his speed increases, and it is easy to find out from the above figures what its mean value during ground-contact must be; for if the constant upward force, equal to body-weight, which is sufficient to support the body standing statically on the ground, becomes a regularly-repeated one, operating, say, for only half the time (as it does in running), then its mean value while it *is* acting must be *twice* body-weight. In the case of the sprinter, for whom the operating time is reduced to the 40 per cent quoted above, the mean value must rise to as much as $2\frac{1}{2}$ times body-weight (100/40). This reasoning is only an expression of the fact that the sum of all the identical impulses delivered by the ground to the runner during this small fraction of his running-time exactly counteracts that due to body-weight acting constantly downward during the whole of this time.

The only external horizontal force acting on the runner on level ground is that due to air resistance—often quoted as about 4 lb wt at sprinting speed in still air—so the mean horizontal component of ground-reaction during contact will be about 10 lb wt for a sprinter moving at constant speed. It must be remembered, though, that this is only the mean, or average, value during the time when the foot is in contact with the ground. It may well be that a foot-placement well ahead of G creates a measurable checking component of force immediately on touch-down, particularly if the runner is over-striding. If this happens, an extra compensatory thrust will be needed in the later stages of the contact phase. Such effects, revealed in force-platform experiments, are, perhaps, inevitable sources of energy-loss, and cause minute changes of the speed of G.

The simple reasoning used above cannot be used to assess the *maximum* values of these components of force, which must be considerably higher than their means; nor can the way the components change during a stride be found without some device such as a force-platform, but they evidently do change as speed increases from rest. At the start, when acceleration is considerable, ground-reaction is forwardly-inclined throughout the contact phase; but, as speed increases, this force becomes more nearly vertical, especially on touch-down, until a speed is reached at which its immediate direction is vertical, and only develops the required forward inclination later. Although this may be regarded as an efficient running-speed, in the sense that no checking loss occurs at any stage, it is not likely to be the highest speed that the runner is capable of. The sprinter, for example, may well tolerate some checking reaction from the ground at touch-down—one that promotes the flexion of a more stiffly-braced leg, as in jumping—so that an increased degree of eccentric contraction gives rise to more positive work done by the leg in the later stages of ground-contact, when its extension again becomes possible.

Optimum running conditions

If by 'steady running' we mean the exact alternate repetition, by arms and legs, of the movement-pattern associated with each stride, then we are led at once to the fundamental relation:

running speed = stride-rate × stride-length

In mechanical terms, 'running-speed' means the mean horizontal speed of G over any complete whole number of strides; and 'stride-length' is the horizontal distance travelled by G while one stride is completed, i.e. between, say, one touchdown and the next one by the opposite foot.

The closed curves of Fig. 35(a) and (b) show the contrasting shapes of the paths taken by the ankles of a sprinter and a distance-runner, respectively, with respect to the runner's centre of mass, the motion being in both cases from left to right. Ground-contact is made when the ankle is at C, and lost when it is at D; the corresponding simultaneous positions of the recovery ankle being at A and B. H shows the general region of the hip, and successive positions of the ankles, obtained from film-analysis, are marked at intervals of $\frac{1}{64}$ s. Each of these components moves once round the curve in every two strides, the movement being in the clockwise sense. The runner is out of contact with the ground while ankle-movement goes from B to C, and, simultaneously, from D to A. Ground-reaction builds up quickly to its maximum when the ankle is nearly underneath G, and falls off as the final 'controlled' effort is made by the extending leg-components when the ankle approaches D.

We know that in steady running the ground-reaction is predominantly vertical, and we now see, from Fig. 35, that its mean value is, in terms of body-weight, the ratio of the time for a complete stride to the time spent in the ground-contact part of it: i.e. the ratio of the times for the ankle to move from C to A and from C to D—2·5 and 2·0 times body-weight in the cases shown. If we assume that these figures represent the greatest mean values capable of being tolerated, in each case, by the runner, then one way to increase running-speed would

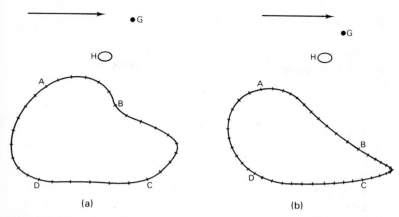

Fig. 35. Typical paths taken by the ankles of (a) a sprinter, and (b) a distance-runner, with respect to the centre of mass, G, of the body as the runners move to the right.

The recovery-ankle moves between A and B while the 'driving' ankle goes from C to D during ground-contact.

Ankle-positions are marked at intervals of $\frac{1}{64}$ second, and the general position of the hips is at H.

be for each stride to be exactly as it is already, but completed in shorter time. This would keep the above time-ratios the same, and the stride lengths would be unaltered, but the stride *rate* would be increased. Experience shows, however, that the rate of expenditure of energy in running increases very rapidly as the speed increases, and this suggests a fairly definite limit to the rate at which the limbs of a sprinter can be moved about. Similarly, there must be an 'economic' stride-rate for the distance-runner, depending on the distance he proposes to run.

If no great increase in stride-rate can be developed by training, then improvement might come from a modification of the 'pattern' of the stride: for the sprinter, perhaps, a faster backward movement of the grounded foot with respect to G at the expense of a slower movement in other parts of the cycle of its motion; for the distance-runner, concerned with energy-conservation, avoidance of energy-wasting changes of speed by all parts of the leg during all parts of the cycle.

Fig. 35(b) shows how the latter condition is fulfilled as far as the extremity (and therefore the whole of the lower part of the leg) of the distance-runner is concerned. Apart from the nearly straight and uniform motion of the grounded phase, there is the smooth recovery from D to A, followed by the leg-straightening drop and the very slow foot-movement when a big change in its direction does have to be made somewhere between B and C.

The sprinter's concern is to maintain maximum forward speed of G, i.e. to develop maximum rate of backward movement of the grounded part of the foot with respect to G between C and D in Fig. 35(a). The best that can be done in this regard is to free the striding movement from time-wasting in any part of its cycle. Leg-movement in this form of running can be understood more easily by using the jointed model of Fig. 34, for this shows how the linear speed of the foot results from a combination of oscillatory 'swinging' motions: one of the thigh about the hip (which is itself moving roughly in phase with it), and the other of the lower part of the leg about the knee. Clearly, these oscillatory motions are never in phase with each other: when, for example, the thigh has reached its extreme forward and rearward positions, and is momentarily at rest, the lower part of the leg is rotating about the knee at nearly its maximum angular velocity—it is near the *centre* of its swing. The motion of the thigh, unlike that of a simple pendulum, tends to be maintained at or near its maximum value for much of the range of its swing. This makes for time-saving and is an essential feature of the 'driving' phase of a stride, for the backward linear speed of the grounded foot must be kept up if G's forward motion is not to be checked. During recovery, the thigh's angular velocity is able to build up rapidly because the effective moment of inertia of the whole leg about the hip is reduced by the close passage of the lower part of the leg under it.

Build-up of foot-speed in sprinting   At or near the extreme forward 'knee-lift' position of the thigh, its retardation to rest and the subsequent reversal of its motion is helped by the accelerated extension of the lower leg—now rotating in such a way as to share, and then rapidly to remove and reverse, the thigh's own angular momentum about G. This is a vital, time-saving part of the stride, for it is here that the simultaneous downward and backward motion of the knee, and the extension of the foot and calf about it, maintain the speed of the foot around the blunt forward protuberance characterising its path in Fig. 35(a). The high knee-lift is important in this final build-up of foot-speed, for it provides time for the necessary change of direction, and puts the ground-contact point, C, about midway between the extremes of the thigh's backward and forward movement: a point at which its angular velocity would be high. This also points to the need, not only for rapid acceleration of the thigh from the high knee-lift, but for extreme flexibility and freedom in its rearward swing towards its final momentary rest-point.

The latter condition is seen to be even more necessary when we realise that if a faster foot-speed has been achieved in the ground-contact phase, this tends to reduce the fraction of the time occupied by this phase, and so possibly increase beyond an acceptable level the mean vertical thrust of the ground on the driving foot. This can only be avoided by keeping in touch with the ground over a greater range of foot-movement, i.e. by increasing the distance between positions C and D of Fig. 35(a).

Development of maximum ground-reaction

When all the factors involved in the attainment of high running-speed are considered, the ability to tolerate a high value of vertical ground-reaction appears as the most fundamental one; for if there are physical limitations to stride-rate and leg-extension, we can still increase stride-length (and therefore running-speed) if we can obtain from the ground a bigger vertical reaction. As seen above, the greater this is, the higher is the proportion of the time spent in the air, and, if the other factors remain constant, the longer is the time available for the recovery leg to build up fast foot-speed for the grounded phase of the stride. The development of this reaction is as important in sprinting as it is in jumping, and demands the same fundamental technique.

The way in which G responds to the vertical thrust of the ground on the driving leg of a sprinter is seen in Fig. 36, which shows, approximately, its changing vertical position as

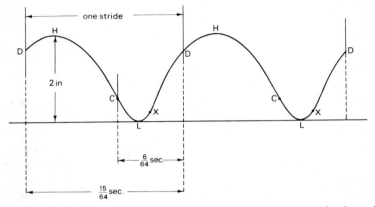

Fig. 36. The undulating motion of G for a sprinter, based on the data of Fig. 35(a), and plotted against time.

The high and low points, H and L, are separated vertically by approximately 2 in; as in Fig. 35(a), the body is in contact with ground between C and D.

Flexion of the grounded leg occurs between C and X; the sharp curvature near L, compared with that near H, is evidence of big vertical ground-reaction between C and X.

each stride progresses. Here, G drops through nearly two inches from its high-point to the level it has when the lead-foot reaches the ground at C, and is then quickly brought to momentary rest at its low-point as it passes over the grounded foot. Continuing ground-reaction then reverses the motion of G, sending it past X, and then at a progressively slower rate, so that it is moving quite slowly when the final ankle-extension drives the body clear of the ground at D. G then moves up to its high-point again, travelling as a body in free fall.

Now the condition that G shall continue to move between the same two levels is that the mechanical work done in bringing it to rest at its low-point—'negative' work done by ground-reaction via the stiffly-braced, resisting, and slightly bending leg—must be equal to the 'positive' work done in driving it up again to its high-point. This work ceases at D, but has given the body sufficient kinetic energy for G to reach the high-point without further aid. The notable contrast between the distances over which these equal amounts of work are done shows how much greater is the force capable of being exerted by a limb resisting flexion, compared with that exerted by it when it is extending rapidly.

Use of the free limbs   The last point needs examining more precisely. Although ground-reaction ceases to perform negative work on the body when G reaches its low-point, there is no reason why the driving leg should not continue to bend for a short time after this and, in doing so, continue to transmit this very big force to the body until, at X (Fig. 36), it is able to start its own extension. This is what the driving leg does; and it does it under conditions already discussed in connection with jumping (p. 146), where the upward acceleration of other parts of the body, particularly the free limbs, is responsible for the delayed extension of the components of the leg.

This happens in running because, when G is near its low-point, not only are the knees passing each other, but the arms are doing so too: all 'free' limbs, which had been moving downwards, are about to move upwards again, either in front of the body or behind it—all are accelerating upwards and therefore imposing on the supporting leg a force of much more than their own weight, and evoking from the ground a correspondingly greater vertical reaction. It is this, of course, that contributes to the rapid change in the direction of G's motion near its low-point; but the effect does not stop there, for the upward limb-acceleration continues for a time as the arms and the free leg start to rise, causing further flexion of the driving leg even as G moves upward with them. No doubt the phenomenon of eccentric contraction operates here, giving the driving leg increased capacity to perform positive work when it is free to extend in the final stages of the drive from the ground.

This is the effect that tends to maintain a high value of ground-reaction for a reasonable part of the stride. It depends for its efficacy on the correct timing of the acceleration of all free limbs, and it is intimately connected with what is called 'good style' in running.

Problems

1. A series of dumb-bells, or similar bodies, of widely differing mass, are suspended, in turn, from the end of a cord hanging from a suitable high point. They are adjusted to be at shoulder-height so that each can be forced at maximum velocity from the shoulder in a horizontal 'putting' action, using full arm-extension but no movement of the shoulder: this must be kept firmly in contact with a rigid barrier (e.g. a wall) behind it.

An observer notes, on a vertical distance-scale, the height, h, through which each of the projected bodies is made to rise before coming to rest momentarily at the extremity of its arc of swing. The work done is measured by the product $m \cdot g \cdot h$, and it can then be used to find how the initial kinetic energy of each body, and the force it has been possible to exert on it, depends on its mass.

The experiment can be varied by allowing a limited shoulder-movement: first in the early stages of the action, so that it ceases too soon; then in the final stages, finishing last.

2. Prove that if the constant force, F, of Fig. 32(a), were to be quadrupled, then the final velocity of the body would be doubled, as would be the impulse of the force; but the time of its action would be halved.

3. Discuss, in terms of energy-transformation in a collision, and considerations of impulse and momentum, the difficulty of kicking a football satisfactorily when (a) it is seriously 'over-inflated' and (b) it is 'under-inflated'.

4. Describe the characteristic hip-and-arm action in race-walking, and explain how it tends to keep the vertical component of ground-reaction nearly constant throughout each stride. Show that G's vertical movement is thereby minimised, and that this helps to keep unbroken contact with the ground.

CHAPTER TEN

Twisting Activities

Basic principles

Twisting movement has already been introduced in connection with the properties of principal axes and the effect of a lack of coincidence between any 'stable' one of them and the body's 'axis of momentum'—the direction of the vector representing its total angular momentum. It was noted that, in free fall, the axis of momentum necessarily retains its constant direction, but that a posture-change, i.e. a redistribution of body-mass, can change the direction of the principal axis formerly coinciding with it, and give rise to a conical motion of this axis about the axis of momentum: a motion known as 'nutation' (Fig. 20). Evidently, this condition results from the fact that the body's angular momentum (still confined to the *axis* of momentum) now has components about more than one principal axis—the body is rotating about more than one such axis at the same time.

A simple way to gain experience of this type of motion is by experimenting with a cylindrical object, such as a metal container, which has a longitudinal principal axis AB (Fig. 37) and transverse ones through *G* in any directions at right-angles to it. This can be projected into the air with rotation about AB and one of its transverse axes by holding it near its

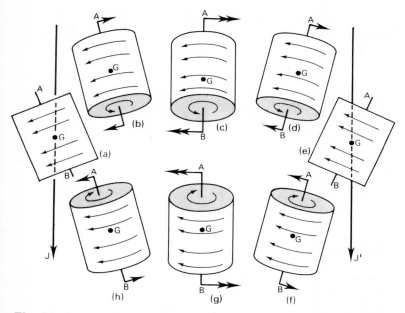

Fig. 37. A sequence of attitudes adopted by a nutating cylinder seen projected at right-angles to the axis of momentum (shown by arrows drawn in the plane of the figure at J and J').

The longitudinal axis, AB, has an apparent oscillatory motion shown by arrows at A and B: arrows on its surface show the cylinder's own rotation about AB.

The *positioning* of the diagrams has no mechanical significance: they are displayed like this to emphasise the conical character of the surface traced out by AB, the cone really having its vertex at G.

lower end with AB inclined somewhat away from the body, as in attitude (c), and discharging it with a spinning action about AB. If the right hand is used, the cylinder will rotate about AB in the sense shown in Fig. 37 and will show changes of attitude (but not changes of position) similar to those shown in the series of diagrams (a) to (h), where the end of the cylinder presented to the observer is shaded.

If observation is made on the same cylinder, it will soon be clear that the cone-angle traced out by AB depends on the relative rates of rotation of the cylinder about the two axes concerned. If the rate of spin about AB is comparatively slow, the motion will look more like a somersaulting movement about the transverse axis; if the longitudinal spin is much faster than it is about the transverse axis, the cone-angle is much smaller. It will also be found that the nutation-rate in the latter case is slower than the rate of rotation of the cylinder itself about AB, especially if the cylinder is rather long and narrow; for a flat, disc-like object, the reverse is more likely to be true. Experiments with a really long and narrow body, such as a broom-handle, will show the difficulty of making it move in anything more than a very rapidly twisting somersault: a straight aerial pirouette is almost impossible.

Application of the 'method of projection'   It is now desirable to show, at least qualitatively, that the phenomena recorded above are consistent with the fundamental conditions under which the cylinder is moving: that its angular momentum about *G* is constant in magnitude and direction—the direction of its axis of momentum. We must first realise that, no matter how the cylinder's attitude changes as the axis AB 'gyrates', at no time can it have resultant angular momentum about any line through *G* at right-angles to its axis of momentum. If, therefore, in Fig. 37, it is being viewed all the time along a line in this 'plane of zero momentum', as we have called it, then the rotary motions shown projected thus in each of the series of diagrams must be self-cancelling as far as angular momentum about this line is concerned.

Examining the 'projected' changes of attitude in Fig. 37 we find two rotary motions going on—the rotation of the cylinder about AB (indicated by the arrows on its surfaces), and the rotation about *G*, first in one sense and then in the other, of the axis AB itself (shown by arrows at A and B). The 'conical' motion of this axis causes its 'side-view' projection to be momentarily at rest in (a) and (e), where no *rotary* motion of the surface of the cylinder is apparent either. At other stages, when one end or the other is being presented to the observer, the two rotations are seen always to be opposite in sense, their contrary components of angular momentum building up together between (a) and (c), diminishing to zero together between (c) and (e), and continuing to change thus in the other phases of the conical motion of the cylinder. Evidently, therefore, the motion observed is a constant cancellation of angular momentum in this plane: furthermore, *along* the axis of momentum the components are additive.

If one examines a cylinder held in the various attitudes shown in Fig. 37 and given slow rotary movements of the type described, it will soon be realised that, in popular terms, rotation about AB would be 'seen' more prominently, and its projected component of angular momentum made more obvious, if AB were to be tilted more nearly into the observer's line of sight instead of being nearly at right-angles to it. At the same time, a *smaller* component of angular momentum due to AB's own rotary movement would be projected in this direction, for the cylinder would appear foreshortened. Clearly, if a wider-angled nutation of the cylinder is required, the balance between the two contrary components of angular momentum will have to be achieved by a slower rate of rotation of the cylinder about AB, as is indeed found to be the case. However, if the cylinder is changed for a longer, narrower one (to which the extended human body closely approximates), its small moment of inertia about AB, compared with that about any *transverse* axis through *G*, will require a much faster rotation-rate about AB to maintain a big enough component of angular momentum just to keep the nutation-angle small—as in the aerial pirouette of Fig. 20(b).

The twisting somersault

Let an extended human body, behaving, as we may suppose, like a cylinder, be given enough time in free fall to make posture-changes capable of altering the direction of its longitudinal principal axis. (Such posture-changes could, conveniently, be rotary movement of the arms in the frontal plane.) Starting with a straight aerial pirouette, the longitudinal axis coinciding exactly with the axis of momentum, let the former be slightly realigned so that a small-angle nutation starts. This has the character of a 'twisting' pirouette, and, from what has been said above, the rate of spin about the longitudinal axis (AB in Fig. 37) will now have been somewhat reduced, although still being much faster than the rate at which AB is itself sweeping out its conical path around the axis of momentum. If we now imagine AB to be deviated progressively further from its original direction, we shall realise that the motion begins to look more like a twisting somersault and less like a twisting pirouette, and that each increase in the cone-angle generated by AB is accompanied by a reduction in the rate of rotation of the body about AB, provided that the body retains its same approximately cylindrical form.

Eventually, when AB has been deviated through nearly a right-angle, the body's motion will be recognised as a true twisting somersault; and when the 'twist' has been taken out of it by a final rotary posture-change, the body will be performing a straight somersault with one of its *transverse* axes now coinciding with the axis of momentum. Necessarily, for somersaulting, this axis is a horizontal one rather than the vertical one suggested here; but it is instructive to reproduce the transition from pirouette to somersault, as described above, or to proceed in the reverse way, using a cylinder held in the hand and made to move with an increasing (or decreasing) nutation-angle from one straight rotary movement to the other. One interesting feature of the transition is that, although the inclination of AB is made to vary, its rate of nutation about the axis of momentum does not change—it is the same as the final somersaulting-rate, no matter what the cone-angle. The effect of the body's slight lack of symmetry about AB (its moments of inertia about CD and EF in Fig. 19 are not quite the same) merely causes AB to 'sway about' somewhat as it traces out its conical surface.

The conical surface is distorted very considerably when, as in high-jumping, major changes of body-posture are made in the air. The effect, then, is to change the cone-angle rapidly as the action goes on: a change associated with alteration in the rate of rotation of the body about AB; and, because of its changing radii of gyration, there will be change in the nutation-rate also.

Reversion of attitude in the air, with zero angular momentum ('catting')

The method of projection whereby a body's motion may be viewed and analysed in specific directions, can be used fruitfully to understand how it is possible for a motivated body, e.g. an animal such as a cat, or an agile gymnast, to turn in the air, from rest, through any necessary angle about its longitudinal axis, and to make a controlled landing.

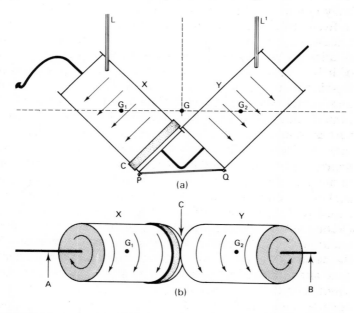

Fig. 38. In (a), the 'cat' is seen along its transverse principal axis through G, the other axes being indicated by broken lines. A collar, C, near the end of the cylinder X, forms a groove around which the rim of Y can roll.

The rotary motions set up in the model by the contracting elastic band, PQ, should be compared with those of Fig. 37.

Pairs of wires are attached at L and L′ to act as 'legs' on which the model may land.

This matter can be studied more easily with a working model like that of Fig. 38. A spoke from a bicycle wheel is bent into the form of a V, the angle between the arms of the V being approximately a right-angle. This is threaded through small holes bored centrally through the ends of two cans, X and Y, to form something capable of representing an inverted body which, on being released in the air, immediately assumes this V-shaped form. To give motive-power to the model, a piece of thin hat-elastic is stretched between holes bored in the rims at P and Q: this contracts when the model is free, just as muscles do, to perform the required movement. A smooth collar, around which the end of Y can roll, is provided by several turns of thick adhesive tape wound carefully near the adjacent end of X, as shown. It is convenient to bend one end of the wire spoke, to indicate any change in the attitude of the V; and some appendages, attached to X and Y to represent legs, will more easily show the rotary movement of these parts when the model turns over.

When the model is held over a table, in the inverted attitude of Fig. 38(a), and then tossed upwards through 1–2 ft, it will turn over in the air and can be caught, or landed on a cushion with its 'legs' pointing downward. It will be noticed, though, that the V lies no longer in a vertical plane: it will have turned somewhat away from it in a sense opposite to the one in which the rest of the model has rotated.

This last fact gives a clue to what has happened. In bringing the points P and Q together, the stretched elastic has not only started X and Y rotating about their common 'bent' axle, but has forced them to move with this axle around the line joining their individual centres of mass, G_1 and G_2. When observed *along* this line, then, the projected view of X and Y (one covering the other) will show exactly the same rotations as seen in Fig. 37(g): rotations responsible for equal and opposite cancelling-components of angular momentum. When viewed as in Fig. 38(a) there is no rotary motion to be seen, and in (b) similar cancelling effects are apparent for both X and Y as, in this projection, they rotate in opposite senses *about* the bent axle and also in opposite senses *with the two parts* of the bent axle as it changes its attitude in accordance with the direction of the arrows shown at A, B, and C.

Clearly, then, it is possible for a flexible body to turn over in the air with no angular momentum about any axis; but there is more than one way of doing it. To produce, simultaneously, the rotary effects described above, an animal or gymnast would develop rapid circumduction of the upper part of the body: a movement which, in reaction, necessarily produces the same conical motion of the rest. In practice, however, an animal often makes use of the fact, remarked on earlier, that a long, slim body has a much smaller moment of inertia about its longitudinal axis than about any transverse axis through G. The animal therefore uses its abdominal muscles to ensure that its fore-parts (Y, in Fig. 38) are first rotated into a good landing-attitude before the rest of the body is so treated. Each rotation turns the whole body through a comparatively small angle about the axis G_1G_2, just as the simultaneous rotations do in the movement discussed earlier.

Other, more laborious methods of turning in the air are available, using rotary arm-movement; but however any such movement may be performed from rest, the resultant angular momentum of the body about any axis through G must at all times be zero.

Problems

1. Project into the air a model of the erect human form, or a cylinder of about the same shape, so that it performs straight somersaults. Now make it perform twisting somersaults by also giving it rotation about its longitudinal axis, AB. Note the very rapid rotation that the model has to have before it becomes obvious that AB is moving in a wide-angled cone and not in the original plane.

Do the same experiment with a short, squat cylinder and explain the different result obtained.

2. A diver leaves the board in a forward dive with arms outstretched at the sides. He then gives the extended arms a frontal rotary movement, to bring the right one over the head, the other to the side. What effect will this have on the longitudinal axis of the body, and in which sense will the body start to twist? If the twist is to be 'killed' before entry by an exact reversal of the arm-movement that started it, point out the need for correct timing.

3. A high-jumper approaches the bar in a convenient direction from the left. His 'drive' from the ground gives him angular momentum in the 'forward' sense about his transverse principal axis, and in the anticlockwise sense about the sagittal and longitudinal axes, as seen from the rear and from above, respectively. Use a model to find, approximately, the directions of these components of angular momentum, and so estimate that of the axis of momentum. Demonstrate, with the model, the 'conical' motion of the longitudinal axis as a satisfactory layout is obtained in bar-clearance.

4. In the flop style of high-jumping, a jumper finds himself arched back over the bar with his body symmetrically disposed on each side of its sagittal plane. Consider whether there should be any twisting at this stage; if not, show that the axis of momentum must have the same direction as the transverse axis of the body, and that this is nearly the same as the direction of the bar. Can these facts be used to define a pure 'flop'?

Index

B

This book is intended for students of physical education and is particularly suitable for those taking the B Ed and PE courses for the Teacher's Certificate. It will also be of interest to athletes, coaches and anyone interested in human body movement.

The purpose of the book is to explain the mechanical principles involved in human movement. It is basically concerned with ways in which mechanical principles are either made use of in human movement or are exemplified by it.

The approach to the subject is almost completely non-mathematical, and appeals to demonstration and experiment to bring out the fundamentals of the subject. At the end of each chapter there are exercises – questions to answer, practical tests to perform, puzzling matters to be thought about and things to argue about.

This book is one of a series in physical education published by Crosby Lockwood Staples. Other titles include: Exercise Physiology by Vaughan Thomas. Curriculum Development in Physical Education edited by John Kane. Sociology of Physical Education.

The Author

Bernard J Hopper was formerly Head of the Science Departmen St. Mary's College of Education, Twickenham. He has lectured or the scientific aspects of human movement at several British universities and in Canada – notably at the University of Guel where he established a course ir the subject. He was a Member of the Association of Training Colle Lecturers and a member of the Examining Board for Physics in London Institute of Education. For many years he has made a study of human movement in athletics, judo and weightlifting and has given numerous lecture to organisations in these fields. Many articles of his have appear in track and field journals both ir the United Kingdom and North America.

Paperback

Crosby Lockwood Staples   London

£2.25p n

Printed in Great Britain   ISBN 0 258 96907 5

DATE DUE